A PROSPEROUS WAY DOWN

Principles and Policies

Howard T. Odum AND
Elisabeth C. Odum

UNIVERSITY PRESS OF COLORADO

Published by the University Press of Colorado
5589 Arapahoe Avenue, Suite 206C
Boulder, Colorado 80303

All rights reserved
First paperback edition 2008
Printed in the United States of America

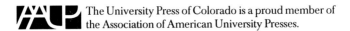 The University Press of Colorado is a proud member of
the Association of American University Presses.

The University Press of Colorado is a cooperative publishing enterprise supported, in part, by Adams State College, Colorado State University, Fort Lewis College, Mesa State College, Metropolitan State College of Denver, University of Colorado, University of Northern Colorado, University of Southern Colorado, and Western State College of Colorado.

∞ The paper used in this publication meets the minimum requirements of the American National Standard for Information Sciences–Permanence of Paper for Printed Library Materials. ANSI Z39.48-1992

Library of Congress Cataloging-in-Publication Data

Odum, Howard T., 1924–2002
 A prosperous way down : principles and policies / Howard T. Odum and Elisabeth C. Odum.
  p. cm.
 Includes bibliographical references and index.
 ISBN 0-87081-610-1 (cloth : alk. paper) – ISBN 978-0-87081-908-7 (pbk. : alk. paper)
  1. Power resources. 2. Economics. 3. Environmental policy. 4. Human ecology. I. Odum, Elisabeth C. II. Title.

TJ163 .O38 2001
333.7–dc21
                                                                    2001020245

17 16 15 14 13 12 11 10 09 08   10 9 8 7 6 5 4 3 2 1

CONTENTS

FIGURES

TABLES

PREFACE

As the global crescendo of information and investments rushes toward the culmination of civilization, most of the six billion people on Earth are oblivious to the turndown ahead. It's time for people to recognize what is happening and how they will be forced by circumstances to adapt to the future.

Studies in many fields in the twentieth century showed how systems and their pulses of growth follow common principles of energy, materials, and information. The following are our previous books on these concepts:

1971 *Environment, Power and Society,* John Wiley, New York, 331 pp. (H. T. Odum).

1976, 1982 *Energy Basis for Man and Nature,* McGraw Hill, New York, 337 pp. (H. T. Odum and E. C. Odum).

1983 *Systems Ecology,* John Wiley, New York, 644 pp. (H. T. Odum).

1993 *Ecological Microcosms,* Springer-Verlag, New York, 555 pp. (R. J. Beyers and H. T. Odum).

1994 *Ecological and General Systems* (reprint of *Systems Ecology*), University Press of Colorado, Niwot, 644 pp. (H. T. Odum).

1995 *Maximum Power,* University Press of Colorado, Niwot, 644 pp. (C.A.S. Hall).

1996 *Environmental Accounting, Emergy and Decision Making,* John Wiley, New York, 370 pp. (H. T. Odum).

1997 *Environment and Society in Florida,* Lewis Publishers, Boca Raton, FL, 499 pp. (H. T. Odum, E. C. Odum, and M. T. Brown).

2000 *Modeling for All Scales,* Academic Press, San Diego, 458 pp.
(H. T. Odum and E. C. Odum).

Now a book is needed to anticipate the future and draw policy recommendations from the principles. This book explains how:

- human society follows general principles;
- growth is but a stage in a resource cycle;
- the policy that works depends on the stage in the cycle;
- the next stage is economic descent; and
- the descent stage can be prosperous.

We know from teaching that many students who hear the story of energy systems causality are profoundly influenced. Now, as signs of downsizing are appearing in our economic life, it is time for everyone to consider fundamental changes in policy and plans.

Prediction or Advocacy. Much of this book is prediction. If the principles are correct (see Chapters 4–8) and we interpret their application correctly, then our recipes for a prosperous future are a prediction of what society will be forced to consider. If civilization is to progress, it has to learn to advocate the patterns that these principles predict. In the process a growth culture will be able to change smoothly into a culture of descent. However, history records many systems that crashed instead. We have tried to confine recommendations to those derived from these principles.

We summarize the present condition and the future views of others in Part I, introduce the concepts and principles of prosperous descent in Part II, and make policy recommendations derived from these policies in Part III. Chapters 9–12 deal with policies to sustain our present civilization during transition; Chapters 13–18 show policies for prosperous descent; and Chapter 19 summarizes.

ACKNOWLEDGMENTS

This book was started while on sabbatical leave at the Lyndon Baines Johnson School of Public Affairs, The University of Texas at Austin. We are grateful for public policy interactions with Bob King, then Director of Natural Resources Division, Texas Department of Agriculture. Suggestions were made by Tom Abel, Joan Breeze, George Darrow, Amy DeHart, Peter Hulm, Daeseok Kang, Mary Logan, Josh Orrell, Morris Trimmer, and many others. Charles A.S. Hall, Syracuse University, critiqued the whole manuscript and suggested adding chapters in Part I. We thank the Worldwatch Institute for the use of data from the Worldwatch Database disk. Joan Breeze was editorial assistant.

A PROSPEROUS
WAY DOWN

I

THE APPROACHING
SUMMIT

The summit of our civilization is just ahead, so we all need to consider how our lives will change and the plans we should make accordingly. Part I considers the present condition, controversies about the future, and the general idea that a natural turndown and descent can be *prosperous*. After Chapter 1 introduces the book, Chapter 2 shows our present condition with recent data, and Chapter 3 reviews the widely different writings of other authors about the future.

Perhaps then readers will be ready for the chapters on systems principles in Part II to explain what is happening. In Part III we use those principles to recommend policies.

1

INTRODUCTION
TO THE WAY DOWN

Like a giant train, the world economy is slowly cresting its trip up the mountain of growth. It may be ready soon for its long trip down to a more sustainable lower level. The developed nations that were leading on the way up are poised for leading again, but this time down. In Chapter 1 we explain the concept of *A Prosperous Way Down,* pointing to the later chapters on our present condition, the views of others, the principles by which the global system is understood, and the policies required for society to adapt. We also explain why we need to think with systems diagrams, the analogies with ecosystems, our summary of public skepticism, and the flip in attitude that is likely. Recall the story for children about the little train going up the mountain (Figure 1.1): "I think I can; I think I can." Then coming down: "I thought I could; I thought I could."

THE PROSPEROUS WAY DOWN

Precedents from ecological systems suggest that the global society can turn down and descend prosperously, reducing assets, population, and unessential baggage while staying in balance with its environmental life-support system. By retaining the information that is most important, a leaner society can reorganize itself and continue making progress. The situation is analogous to the human brain, which regularly dumps less essential information in short-term memory while gathering what is important for the long-term memory.

The reason for descent is that the available resources on Earth are decreasing. Each year more effort is needed to provide the fuels, water, wood, fish, soil, food, electric power, and minerals on which everything else is based. More and more of the economy goes into concentrating what remains with less left for the private lives of people. More

Figure 1.1. The Story of the Little Train (Ann Odum).

and more of the resources supporting the developed nations are diverted from people in other countries by the global economy. The present levels of our urban civilization cannot be sustained indefinitely on the worldwide declining concentrations of resources (see Chapter 10).

Make no mistake, this is *not* a proposal for less growth. It is recognition that general systems principles of energy, matter, and information are operating to force society into a different stage in a long-range cycle. One set of policies is needed for the transition and another set for the descent. We can also look way ahead at a lower energy period when environmental resources accumulate again.

In spite of tendencies toward economic competition, global cooperation has increased. Global unity was improved by teaching ideals of mutual respect and equitable trade. Resurgence of local characteristics, customs, and environmental adaptation has also occurred, helping people to find a smaller group identity in a large complex world. Strengthening local culture is desirable, provided it is accompanied by mutual respect and shared belief in cooperative working relationships among those who are different. The global significance of the 1999 Kosovo war in Yugoslavia was to establish the principle that military aggression against people who are different is no longer acceptable to the majority of nations.

That the way down can be prosperous is the exciting viewpoint whose time has come. Descent is a new frontier to approach with zeal. The goal is to keep the economy adapted to its global biophysical basis. We have to abandon some of our useless diversions. If everyone understands the necessity of the whole society adapting to less, then society can pull together with a common mission to select what is essential. Presidents, governors, and local leaders can explain the problem and lead society in a shared mission. Millions of people the world over, if they see the opportunity, can be united in the common quest for a prosperous way down. The alternative is a world of selfish battles for whatever resources remain.

UNIQUE BASES FOR EXPLANATIONS

This book is different because its explanations about society come from the general scientific concepts that apply to *any* system. These concepts suggest the constraints on the future to which human society will have to fit. In the language of general science, the system of human society and its environment is self-organizing. Through the initiatives of millions of people all sorts of things are tried daily. Those that work are copied by others and become part of the mainstream system of society.

The processes of nature also self-organize with restless testing. The most familiar example is natural selection among species, but reinforcement of what works also occurs in other kinds of systems and on different scales. Thus the global pattern of humanity and nature is a combination of the stormy atmosphere, swirling ocean, slowly cycling Earth, life cycles of living organisms, ecological adaptations, and the complex actions of human society and its economics.

Theory and research now suggest that many, if not all, of the systems of the planet (and the universe) have common properties, organize in similar ways, have similar oscillations over time, have similar patterns spatially, and operate within universal energy laws. If so, it is possible to use these principles in advance to select policies that will succeed. In other words, humans can use their intelligence and social institutions to avoid some of the wasteful mistakes caused by trial and error, doing a better job at evolving a prosperous world within the constraints of nature.

Unfortunately, we have no procedures for proving that principles are general except to keep testing them in new situations. When a principle is successful in explaining outcomes for many examples, it begins to be more and more trustworthy. The general principles offered in Part II have been applied widely, and evidence has come from many different disciplines. Hopefully readers will recognize examples from their own experiences.

Views and Scales

In Chapter 3 we review the wide range of ideas of other authors about our society. Many of their views are consistent with principles in Part II. What we offer is a way to tell which of the myriad of scenarios from futuristic imaginations are appropriate and likely for the times ahead.

Some of the authors try to find causes in short-term, small-scale processes and mechanisms such as: interactions of economic markets, cultural reactions, global capitalism, national policies, atmospheric changes, religious movements, local wars, technological innovations, and so forth. But the general systems view is that the larger-scale pattern selects what is workable from the trials and errors of the smaller scale. The regime prevails because it maximizes the performance possible for those conditions.

Often implied is that humans can select whatever destiny is desirable—a half-truth. The new hope of our time is that the designs in society that will ultimately prevail can be found more rationally by using large-scale principles more, wasteful trial and error less. The new global sharing of information and ideas makes it possible for billions to learn about world pulsing, and to embrace a new faith that coming down is OK.

In this book we recognize the way the important controls on any phenomena come from the next larger scale, determining the main cycles of growth, turndown, catastrophes, and regimes of energy and material to which society must fit. This is a type of scientific determinism. The paradox is that most scientists restrict their deterministic beliefs to the realms of their specialties. When it comes to society and politics, many share the public's view and deny that large-scale principles control phenomena.

Emergy Evaluations

Many futurists write of processes and change qualitatively, although economic data are sometimes cited. In this book we use a new measure—*emergy*—to evaluate the main inputs, products, and accomplishments of our world on a common basis. It is a special measure of the previous work done to make something, whether the work was done by natural processes or by humans.[1] For example, emergy values of exchanges explain why well-meaning international investments and loans have been crushing underdeveloped countries.

Ecosystem Analogy

Forests, lakes, grasslands, coral reefs, sea bottoms, and so forth are ecological systems (ecosystems). They operate on a smaller, faster scale than civilizations, and humans can more easily see the essence of their complexity in relation to the controlling principles of energy, materials, and information. Like civilizations, they have growth cycles, periods of weed-like growth, and periods of high complexity and diversity analogous to human pluralistic societies. Ecologists have a range of views. Those at one extreme see many random processes and seething interactions of species struggling for existence. Those with our view see a high degree of self-organization involving causal interactions through intermittent pathways best generalized with energy systems principles.

Important for our purpose in this book, many ecosystems grow and decline in cycles that are repeating and sustainable. For example, lake ecosystems have daily and seasonal cycles. Forests have cycles involving many years each. H. K. Okruszko[2] named the stage of peatland decrease as *decession,* the opposite of *succession,* the development stage. The normal cycle of some ecosystems includes sharp "destructive events" like fire or consumer epidemics, which are beneficial in the long run, because they accelerate downsizing to the next stage. Dynamiting old buildings for urban renewal is analogous to the ecosystems' destructive events. Thus we use ecosystem comparisons for insight into the larger-scale cycles of our own society.

Network Diagrams for Understanding

Although the call for a systems view is widespread, most people discuss the problems and solutions with verbal concepts that don't give the mind an understanding of connections. Often people won't take the time to study network diagrams that are necessary to visualize causes. The late economist Kenneth Boulding, a brilliant writer, reviewed our earlier book *Energy Basis for Man and Nature* and wrote that it was not necessary to look at the diagrams. But understanding systems requires a language that shows how the connections work. For an overview of the complex system of humanity and environment, the human mind needs the comprehension that comes from seeing the connected functions of the network simultaneously in the mind's eye.

For human understanding the network first needs to be simplified by aggregating the complexity into the main process and parts that are important. Getting the system view in mind helps in understanding the way structure is related to function. You can see parts, wholes, and consequences at the same time, carrying a systems image in memory.[3] Since basic mini-model configurations apply to different kinds of systems on all scales, a person accumulates ways of transferring understanding to new situations.

Policy from Mini-models

Many—if not most—people trained in science learn about separate parts and relationships, expecting computers to synthesize what the combinations will do. But carrying a simple mini-model of a system in mind is a different methodology from expecting computer simulation of large complex models to generate something of which the mind understands only a part at a time. Policy about complex systems is usually made with whatever synthesis word-models provide. Better policies can result if simple mini-model diagrams are kept at hand to visualize causes.

Scale of View

The human mind is like the zoom microscope, able to change focus rapidly from small scale to large scale. For example, some writers describing the behavior of society as a whole use concepts and language from the smaller scale of human psychology about the behavior of individuals. Sometimes authors use analogies to clarify a point. The authors may mean that the society is the sum of the individual psychological actions. Or the writers may mean that individuals and the society are both examples of the same general systems model. Because words are so all-encompassing with so many alternate meanings, they are not very rigorous for representing systems relationships and many scales.

In Part II of this book we use network mini-models to make points about transition and turndown. Our explanations of how the Earth's economic system works can be best understood by putting the pictorial images of systems relationships in mind.

CONTEMPORARY CHANGE

The summit for the global economy ahead is hidden by the surge of affluence in the wealthy sectors of a few countries. But downsizing is already occurring in many parts of the system. This is the start of the long process of reorganizing to form a lesser economy on renewable resources. If we do not understand the principles that are causing the decreases, we won't plan the needed changes. Without a collective mission to adapt, we are more likely to stumble with delay, failures, fear, desperation, conflict, malaise, pestilence, environmental destruction, and collapse.

The Present Reality

Whether the crest in the United States has been reached yet is not clear because short-term fluctuations of the economy mask long-range trends. The annual increases in gross economic product show money circulating more rapidly. Much of it, however, is through finance and stock markets, and circulates without producing real wealth. There are surges in computers and communication but pathological waste of resources in, for example, excess cars. Other measures show important parts of the economy and Earth systems in decline. Recent books on the future and its policies are wildly different. Some warn of crash and others of perpetual boom ahead.

Trends

In Chapter 2, recent trends in resources, population, information, human welfare, and economic states are quoted from various authors and sources. In Chapter 3 (especially pp. 50–53) many of the authors cited look to the future by extending the trend lines on these indices

Cascari © *1997, courtesy* Vero Beach Press Journal.

of society, usually with properties growing upward. Yet all who know about the causal connections between energy, materials, and growth expect an eventual turndown. The question is when. What is argued is: how many usable fuel and mineral resources are still to be discovered underground? And how much of the present world economy could be supported on the proposed alternative energy sources, most of which have been under intensive research for a half century? Whether turndown is near or to follow later, task forces are needed at local, national, and international levels to plan for transition.

Instead of planning for descent, many writers, journalists, and political leaders encourage a continuation of the established public mind-set on growth that was okay for the time of expanding resource use. For some it is failings in their education; for others it is overfocus on the short range. Nearly six billion people are in denial, and for leaders to speak of a nongrowth period is viewed as political suicide. But the paradigm of growth is a shared global attitude that may switch all at once for all together when the truth becomes obvious through some galvanizing event. Or perspectives may shift gradually as books like this one circulate.

PUBLIC PERCEPTION

Interruptions in fuel supply in the 1970s gave people a momentary glimpse of a resource-limited future. As we cite in Chapter 3, many authors considered how to adapt to lower energy availability. But decreasing before you need to is contrary to fundamental energy principles, as we will explain. In the 1980s the world could be and was still engaged in growth. Plans for descent seemed nutty. When the first draft of this book was written in 1982, "coming down" was considered only by a few as a pleasant, alternative lifestyle to seek as a matter of choice. Publishers did not think their readers would be interested. By the end of the century some decreases began. Some downsizing was erratic, divisive, and competitive, a bitter contrast to the ideal of a prosperous descent.

Not enough people understand the large-scale changes requiring them to change individually. Few have been trained to think about resource limits on the large scale. Few people now believe that principles other than that of the free market controls the overall economy. In the late 1990s the real wealth per person was oscillating even though leaders were still talking about more growth. Inequity, blame, and class consciousness threatened the fabric of society. Many returned to the ways of the nineteenth century, when there was more selfish individualism and competition. Although political pressure to downsize has been directed at government, more—not less—government coordination may be needed to adapt society to the new stages ahead.

Some of the indices of our society (see Chapter 2) had stopped growing by 2001. Perhaps going into the twenty-first century people are more open to explanations of the root causes for change. Many are not happy, blaming others or fostering greed in the economic system. They may be ready for the concepts and policies given here that can make the inevitable descent better.

One New York publisher explained why a trade book on future policies based on energetics and systems principles probably would not sell. He said "people don't believe scientists have any special insight on the future." They don't believe humans, economy, and environment follow collective scientific principles. Especially where people are raised with an emphasis on human freedom and choices, the public does not feel controlled. Many have faith in free market economics, because explosive capitalism fits the stage of weedy growth that has lasted for two centuries.

An important quality of our social species is the ability to reprogram ideals and objectives when it becomes apparent to the majority that it is necessary. When growth is possible, then it is necessary, and everything that goes with the exploitation and competition of expansion stages is regarded as good. Then when adapting to descent is necessary, everything that goes with making that stage efficient becomes good. We even slant history with ideals of the present. People already write about the fanatic, zealous, and sometimes ruthless exploitation used for expansion in the nineteenth century as evil, but it was not the public view then. Exploiters were heroes. It is fascinating that changes in attitudes appropriate for a time of leveling and transition, such as complexity, cooperation, diversity, and environmental adaptation, are already being recognized as new ideals.

According to one principle, systems help maximize their performance by the accumulation of stores of materials, energy, or information, to be followed later by a sharp pulse of growth by a using consumer. This mechanism of change applies to public opinion too. Need for a change and consciousness of it accumulate bit by bit in more and more people until a threshold is reached when the whole group discusses and switches attitude, using the energy from the unified focus to change institutions. Perhaps we are now in the stage of accumulating new attitudes for turndown and descent.[4]

Many books try to enlist people in social movements with the assumption that change depends on human choice. But it may be vice versa, that social change is set by events in the resource-civilization cycle. If readers will stay with us long enough to consider the principles (see Chapters 4–8), they may be open to the predictions and policies that might otherwise seem radical.

2

THE PRESENT CONDITION

Late in the twentieth century, world civilization changed in ways never seen before. New information networks, unexplained economic surges, and controversies abounded regarding resources, population, and global pluralism. This chapter examines factual data on the present condition of humanity and nature in the United States and the world.

As Figure 2.1 suggests, people were beginning to think globally and study world indices to understand the basis for their economy. Figures 2.2–2.14 show some of these indices since 1950. These are the data people use to consider trends and anticipate the future. In 2000, television news and commentary were inundating everyone with images of growth prosperity and assurances that it would continue. Others used the same data to predict a leveling to growth or a turndown and descent. We postpone our interpretations to later chapters after we have introduced the general system principles.

CLIMATE CHANGE

Figure 2.2a shows the oscillating record of carbon dioxide in the air late in the twentieth century. The concentration falls in the summer when carbon dioxide is consumed by photosynthesis of the vegetation and rises in the winter when there is mostly respiration of life and the consumption by fires of industry. But each year the carbon dioxide rises more than it falls because of excess consumption of fossil fuels and cutting of forests. Atmospheric carbon dioxide increased from 290 to 367 parts per million since the start of the past century.

Carbon dioxide is like the glass on a greenhouse holding in heat. The public press spread as fact the assumption that this greenhouse heat is rapidly raising the temperature of the Earth and raising sea level. However, the temperature record for the whole atmosphere

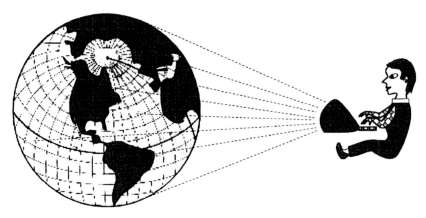

Figure 2.1. Humanity serves the Earth with the power of shared information from global sensing.

viewed from space (Figure 2.2c) did not show much temperature increase.[1] Temperature data available since 1760 suggest that the Earth's temperature mainly varies with activity within the sun, the source of most of the Earth's heat. The sea level has risen only a few inches, and most of that was early in the century before carbon dioxide levels had risen much. J. Oerlemans (1994) showed retreat of seven glaciers early in the twentieth century, but the graphs show that that retreat stopped in 1974. As of year 2000, those in Glacier Bay, Alaska, stopped retreating. Sea ice is decreasing, but this doesn't affect the sea level.

What, then, is happening to the greenhouse heat from increased carbon dioxide? When the temperature of the tropical sea is increased, a little more water is evaporated, absorbing the heat energy into water vapor (see Chapter 15). Later the energy of water vapor makes larger storms that return water as rain and snow, while releasing the heat to the top of the atmosphere, where it goes into space. Larger storms also cause longer dry periods between storms. Data from weather stations on land show a slight air temperature increase (Figure 2.2b), which may be due to the longer periods of drought and clear skies.[2] Slightly warmer seas with more precipitation tend to melt the sea ice and the low altitude part of glaciers while adding more snow and ice on top of glaciers.

HUMAN POPULATION

Nearly everyone knows about the rapid growth of world population, which reached six billion in the year 2000 (Figure 2.3). In a report titled *The Human Race Slows to a Crawl,* the International Institute for Applied Systems Analysis considered the distribution of people by age and extrapolated world population ahead. They found that, at

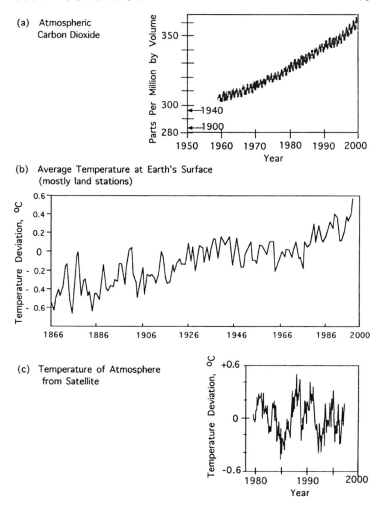

Figure 2.2. Recent atmospheric changes. (a) Record of carbon dioxide in the air; (b) average temperature at the Earth's surface expressed as deviations from 14°C, Worldwatch Institute; (c) satellite measurements of atmospheric temperature expressed as deviations from the mean temperature 1979–1996.

present birth rates, the population would reach about ten billion people, and start decreasing about 2070.[3] Birth rates are decreasing, however, which means a peak will occur sooner.[4] The rapid spread of the AIDS epidemic in Africa may decrease populations there.

ECONOMY

The economy as measured by the gross world product was still growing at the end of the century (Figure 2.4). How much increase in real

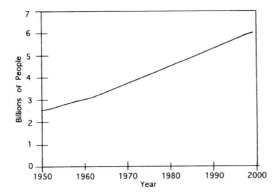

Figure 2.3. Growth of world population.

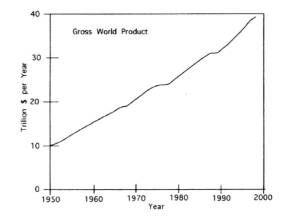

Figure 2.4. Growth of world economic product, expressed in 1997 dollars.

wealth this represents is controversial. As people move to the cities, the circulation of money increases. Some of the increase is for purchases of rural products that people previously obtained without much circulation of money. Some of the increased circulation of money is in rapid transactions of high finance such as buying and selling on the stock market, but the incomes of most people stopped increasing.

The gross economic product of the United States in 1997 was 8.1 trillion dollars per year,[5] 21 percent of the world product, a lower percentage than in the years after the Second World War.

Environmental Disasters

As world populations and economic developments increased, people spread into areas at risk of impact by volcanoes, earthquakes, floods, and hurricanes. C. Flavin and O. Tunali (1996) showed that economic losses due to disasters had increased by nearly ten times late in the century. Between 1991 and 1995 sixteen events cost more than three billion dollars each.

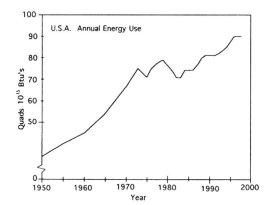

Figure 2.5. Annual energy use in the United States (quadrillion British thermal units).

NONRENEWABLE RESOURCES

At the end of the twentieth century, growth of the economy and population was largely based on nonrenewable resources, fossil fuels, and mined materials.[6]

Fossil Fuel Energy

The industrial revolution started with nineteenth-century coal-burning steam engines. Coal had replaced wood as the main fuel; later oil replaced coal; and late in the twentieth century natural gas was replacing oil. At the end of the twentieth century much of the urban civilization of the world was based on fossil fuels (hydrocarbons: coal, oil, and natural gas). Since these fuels are not being formed by the Earth nearly so fast as they are being used, they are called nonrenewable resources. When fuels became scarce on world markets during the oil embargo of the 1970s, fuel prices went up sharply. In such times money was diverted from other parts of the economy to pay for fuels, and the economy slowed down. Since more money circulated for the same production of real wealth, a surge of inflation resulted.

By 2000, U.S. oil reserves available at reasonable cost were nearly exhausted, and more oil was imported. However, the United States diversified its uses of fossil fuel to include coal and natural gas, as well as nuclear plants, for electric power. Figure 2.5 shows energy use in the United States. The United States represents about 26 percent of the world's energy consumption. A nation's energy use is an indicator of its global influence.

The growth curve of world economic product is steep (Figure 2.4) compared to the curve for world energy consumption, which decreased its rate of growth late in the last century. By one interpretation, the ratio of energy use to dollars decreased because the economy became more efficient. Some even use these curves to claim that energy availability is

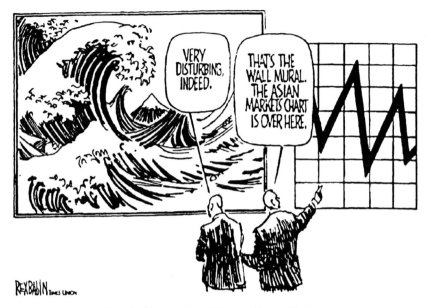

Rex Babin, courtesy Albany Times Union.

unimportant to the economy. By an alternative interpretation, the ratio of money to energy increased because more money circulated for the same real wealth. More money circulates because of increased urbanization and financial transactions.

In 2000 the world consumption of fuel and electricity was still increasing (Figure 2.6),[7] although many of the deposits of the world were becoming exhausted. The economic slowdown in Asia reduced world fuel consumption, and the world prices decreased. Lower fuel prices increased fuel consumption by the United States (Figure 2.5) and helped stimulate its economy in 1998. World fuel prices increased again in late 2000.

Many people knowledgeable about energy and national economies in 2001 were concerned about an approaching time of less-available fuel and permanent rises in world prices, causing the world economy to shrink. Campbell (1997) evaluated the known reserves of fossil fuel and extrapolated the graph of world fuel production into the future (Figure 2.7). A peak in hydrocarbon production was predicted by the year 2009, after which a rise in fuel prices and downsized economy can be expected. The increase in gas shown includes natural gas and gas made from other hydrocarbon fuels.

During the past century there was a shift in forms of energy used to those that released less carbon dioxide. The CO_2 released per unit

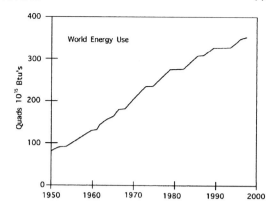

Figure 2.6. World energy use in quads (quadrillion British thermal units). Modified from C. Flavin and S. Dunn 1999.

of energy consumption decreased by 36 percent as the world shifted from wood and coal to oil, natural gas, and nuclear energy–high-quality sources with more energy and less carbon.[8] The increased use of natural gas stimulated the economy more than less-concentrated forms of energy and affected the climate less.

By the year 2000 many resource scientists, new books, and Internet discussion groups and web sites (example: www.dieoff.org) were warning of the economy cresting in the next decade.[9] Journalists and the public once again felt concern over rising prices for natural gas and oil products and shortages of electric power in late 2000. Willard Fey and Ann Lam called for a Manhattan Style Project (Ecocosm) to adapt to the lower energy future.

Materials

The global use of materials increased as more and more countries expanded their populations and economies (Figure 2.8). Main infrastructures were in place in the United States and other developed countries, many heavy industries moved to developing countries, and efficiencies in use and reuse of materials increased. Annual use of materials stopped growing.[10]

In the twentieth century, prices of nonfuel materials decreased even though many deposits were being depleted. Using cheap energy directly and indirectly increased mining efficiency. More and more of the materials came from undeveloped countries in which labor costs were low. The low prices delayed measures to conserve materials and reduce wastes. Although recycling and reuse increased, the generation of solid wastes increased in the United States and elsewhere.

RENEWABLE RESOURCES

Renewable resources are the inputs to the economy from the environment that are renewed by nature about as fast as we use them. Radiant

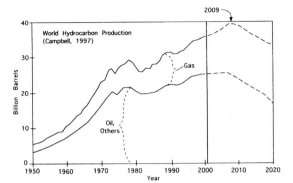

Figure 2.7. World fossil fuel production estimated by C. J. Campbell (1997) for the past and future.

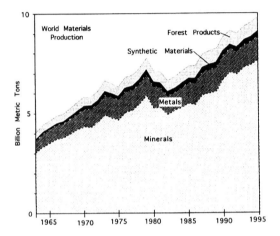

Figure 2.8. World materials production, Worldwatch Institute.

energy from the sun heats the lands and oceans, and this energy drives the atmospheric winds. The ocean winds send waves to the beach shores. The sun's energy varies about 10 percent in fluctuations and cycles. Solar energy drives the cycle of water, producing the rains and snows that continuously renew the freshwaters of rivers and lakes on which civilization depends. Many developments depend on groundwater, the water stored in small spaces in the ground. Winds, tides, and river waters drive ocean currents. Energy deep in the Earth renews the land with geological cycles.

Agrarian human civilizations were based on renewable resources for the thousands of years of prehistory and history before the industrial revolution. But by 2000 agrarian ways of life were mostly gone, replaced by an urban civilization running mostly on fossil fuels. However, the urban civilization also requires renewable resources, waters, lands, and environmental products. Late in the last century fewer flows of renewable resources were left to support further urban growth.

From EARTHTOONS by Stan Eales. Copyright © 1991 by Stan Eales. By permission of Warner Books Inc.

Slowly Renewed Resources

In the surge of population and economic growth, some resources that were being renewed enough for human needs when our civilization was smaller are now being depleted. For example, some groundwaters are being used faster than they are being replenished. Figure 2.9 shows the decreased water available for crop irrigation relative to the number of people.

When fewer people were using the land, peat and soils were formed by ecosystems as quickly as they were used up. People moved from one area to another, giving ecosystems time to restore fertility. Now with dense populations, the peat and soils are being used faster than they are being restored.

Agriculture

Technology, fertilizers, and pesticides increased each acre's agricultural productivity earlier in the last century. But by 2000, there were diminishing returns. The production of grain relative to the number of people reached a maximum (Figure 2.9a). Croplands were lost to salt accumulations (24 percent of irrigated lands damaged) and urban development.[11] Global stockpiles of food diminished to a few months' supply.

Freshwaters

In areas of drier climate, shortages of clean freshwater limited populations and economic developments. Groundwater storages that required many years to accumulate were used up to expand agriculture and urban developments. Irrigated farmlands did not keep up with population growth. The irrigated farm area per person stopped growing

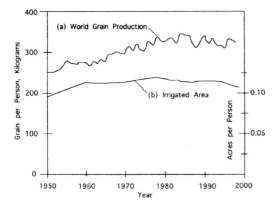

Figure 2.9. Indices of world agricultural production per person per year. (a) Grain production; (b) irrigated land area.

(Figure 2.9b). Cities in dry areas used cheap fuels to distill freshwater from salt waters.

Forests

Forests are slowly renewable where seeding for regrowth is available after harvest. In the last century tropical forests were cleared, but forests were reestablished in some areas in developed countries. Overall, forests of the world were cleared faster than they were replaced. More than half of the old-growth forests were harvested. Much of the clear-cutting converted land use to pasture or urban development, eliminating the areas for forest regrowth. World wood consumption increased (3.5 billion cubic meters by 1990). By 1995, 29 percent of the original forest area had been converted into other land uses.[12]

Fisheries

Fisheries are renewable when the stocks are lightly harvested, leaving populations to reproduce and restore stocks in time for the next harvest. But overfishing reduced stocks so that the world yield from fisheries is the first of the renewable resources that seems to have reached its limit (Figure 2.10). The price of fish in developed countries now exceeds the price of beef. High prices are encouraging more aquaculture such as growing salmon in floating cages.

Biodiversity

The many species of the land and water accumulated during a billion years of evolution and ecological organization. Species take thousands of years to develop through natural processes. On the scale of time of human civilizations, biodiversity is a nonrenewable resource. The destruction of forests, prairies, deserts, and underwater reefs caused extensive extinction of species late in the twentieth century. Of 64 million acres of coral reefs, an estimated 58 percent were damaged or at risk. Global concern over biodiversity loss caused 170 environmental treaties to be developed by 1990.[13]

ENERGY TECHNOLOGY FOR RENEWABLE RESOURCES

The diversion of renewable energy from traditional uses to substitute for fuels and electric power through new technology is highly controversial. Here we report the status of these energy uses, postponing the discussion to Chapter 10.

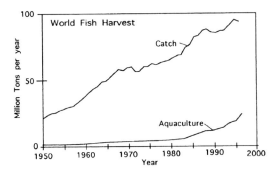

Figure 2.10. World fish harvest.

Solar Technology

Late in the twentieth century there was increased use of solar voltaic cells (cells that generate electricity) for many special purposes. But in spite of sixty years of research, solar technology was not competing with other sources of power. Solar voltaics contributed only 0.2 percent of the electric power in the United States. Prices stopped decreasing. There was little increase in solar water heating. But many like those at Worldwatch wrote that solar technology could replace fossil fuels.[14]

Wind Technology

Windmills for mechanical work such as pumping water and wind-driven sailing ships were important before the industrial revolution. But many wind regimes are not steady enough to favor wind power. Wind technology for electric power increased in the 1990s, supplying 0.88 percent in the United States.[15] (See Chapter 10.)

Hydroelectric Power

Much of the energy of elevated waters in mountains was diverted to generate hydroelectric power, reducing the ability of the river floods to fertilize wetlands and distributaries downstream. Few situations remain for further hydroelectric development that does not interfere with other established river uses. In the United States hydropower supplied 4.4 percent of electrical power used.[16]

HUMAN DIVERSITY

Some of the great diversity of humanity is genetic and inherited. For example, the United Nations recognizes many races. There is also great human diversity in the information of cultural inheritance learned by children as they grow up.

Increased Human Migration

In the twentieth century, migration of peoples increased, facilitated by the easy transportation possible with the fossil fuels, often driven by overpopulation. In the century there were more wars with higher levels of destruction that displaced more people than in any previous century. For example, refugees fleeing persecution increased from 2 million in 1960 to 23 million in 1993.[17] The population diversity increased in most countries.

Diversity of Human Genetics

The great diversity of genetic inheritance in human populations of the world was increasingly recognized as a contribution to global productivity. Genetic diversity helps keep humanity adapted to its environment and less susceptible to disease epidemics. Respect for human biological diversity increased in many countries at the end of the twentieth century.

Diversity of Human Occupations

Human occupations are analogous to the species of an ecosystem. Both refer to the specialties by which a system's work is accomplished. Late in the twentieth century a great increase occurred in the variety of occupations, with more specialization and division of labor in the economy. For example, 340 occupations were found in 1,000 listings in the telephone directory. This is three times a similar count made at mid-century.[18]

AUTOMOBILES AND HIGHWAYS

A major part of the energy and materials of the twentieth-century economy went into increasing use of automobiles, trucks, and highways. The automobile culture that started with mass production manufacture in the United States had spread to the world near the end of the century (Figure 2.11). World bicycle use also increased.[19]

MILITARY EXPENDITURES

In the twentieth century many wars and armed conflicts took place, as well as a cold war between the Soviet Union and the Western countries. World military expenditures and sales of arms increased until the mid-1980s and then decreased after the end of the cold war.[20]

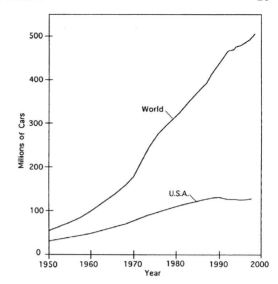

*Figure 2.11. Number of pas-
senger cars (updated from
Herendeen 1998).*

INFORMATION AND ITS TECHNOLOGY

Most would agree that the greatest change at the end of the twentieth
century was the spread of technology for processing and transmitting
information. The world became united by exposure to television, often including the same news, advertising of globally distributed products, sports, music, and drama. Television's influence continues to increase. Use of paper increased as more information was saved in the
traditionally secure form.

Starting in the developed countries, personal computers became
prevalent at home and at work. E-mail and use of the Internet spread
worldwide, decentralizing the spread of information formerly limited
to centers of education and research. A huge increase in short-term
information processing occurred while the growth of traditional universities, research institutes, and information stored in libraries was
slight. The steep growth of wireless phones (Figure 2.12) and the Internet
is an example of the huge increase in communication technology.[21]

ELECTRIC POWER

The proportion of fuels and other energy inputs directed into electric
power increased as part of the greater use of electric technology for
industries, households, and especially for computers and communications. Percent of energy use as electric power was greater in more developed countries. In the United States electric power use continued
its increase, although the use per person stopped growing.[22] Outside
investments accelerated the damming of rivers in undeveloped countries

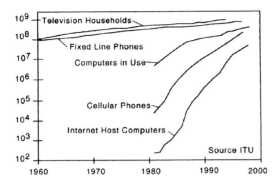

Figure 2.12. Growth in information technology.

for hydroelectric power to stimulate economic development. These projects displaced environmental production. At mid–twentieth century the world's electricity was increased by building nuclear power plants. After serious accidents, construction of nuclear plants was stopped. Accumulation of intensely radioactive wastes from existing plants increased rapidly (ninety thousand tons by 1990).[23]

FINANCE AND INTERNATIONAL EXCHANGE

At the end of the twentieth century there was a sustained surge in the stock market of the United States as a large percentage of the United States population started investing their savings in stocks and bonds. Frenzied buying and selling, including speculation on the Internet, increased the supply of circulating money in finance. Unearned income (income not requiring work) increased. Money was attracted away from investments in other countries. Speculative buying pushed up prices of stocks involving computers and the Internet, even when no dividends were paid. Many professionals warned of the danger of a stock market crash causing deflation (disappearance) of the money that had been created by speculation.[24]

Late in the century capitalism developed a global character as multinational business firms invested in many different countries according to opportunities for profit. International trade increased and national barriers to trade were reduced. Many manufacturing plants in developed countries relocated in less-developed countries where labor costs were lower and environmental regulations less stringent.

Investments by agents of global capitalism in underdeveloped countries rapidly transported mineral, forestry, fishery, and agricultural resources to developed nations. Exports increased to 3.5 trillion dollars by 1990.[25] Third-world debt increased rapidly in the latter years.[26] Policies of lending money to undeveloped countries caused wealth to follow interest payments to enrich the developed nations.

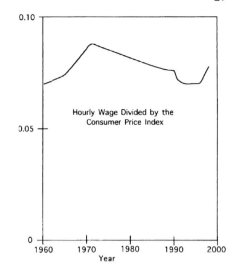

*Figure 2.13. Index of well-being
in the United States calculated by
dividing the annual hourly wage
by the consumer price index.*

Increased populations and decreased welfare in less-developed countries pressured people to migrate to developed nations.

INDIVIDUAL WELL-BEING

Late in the twentieth century, world populations, the world economic
product, and world fuel use grew at similar rates (Figures 2.3, 2.4, and
2.6). But there were huge differences between these indices for different countries. Income per person in the United States in 1995 was
similar to that of European countries and Japan. However, energy use
in the United States was twice as high. Incomes and energy uses in the
United States were ten to fifty times more than those of the less developed nations. After the oil embargo in 1973, wages in the United States
decreased in comparison to consumer prices, data indicating if anything a decrease in standard of living (Figure 2.13).

Income expressed in constant dollars did not increase for most
people within the United States (Figure 2.14). However, more income
for the rich increased the spread between rich and poor.

SUMMARY

In the late twentieth century the Earth system experienced growth in
population, money circulation, energy use, carbon dioxide, electric-
powered technology of communication and computation, international
trade, occupational diversity, pluralistic respect, and global capitalism. It was a century of resource exploitation, with declining availability of renewable and nonrenewable resources. This raised questions
about capacity for sustaining the present levels of civilization in the

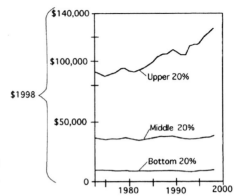

Figure 2.14. Comparison of income in the United States (1998 dollars).

future. In 1998, a wave of concentrated wealth flowed from other countries into the United States. Great differences in well-being developed between rich and poor nations and between rich and poor within the United States.

3

INTELLECTUAL VIEWS
OF THE FUTURE

The voices of intellectuals studying the workings of society are partly obscured from the public by the thundering monopoly of the airways by celebrities, political leaders, advertisers, humorists, and journalists engaged with the short term. But when we look at what has been written, including current bestsellers, we find many authors who write of a turndown ahead. Others see or search for a sustainable steady state. Some write about changes necessary to forestall resource shortages, while others predict major changes with little mention of limits. The warnings span a half century, which is one reason they are being ignored like the boy who cried "wolf." In a counter-reaction, others write of a boom of continued expansion ahead.

In this chapter we sample the futurist literature, quoting some of the authors' concerns and solutions, arranged under ten headings. Obviously the classification is a bit artificial, because the writings of many of the authors could fit under more than one heading.

The first four headings are about turndowns and their causes:

- Warnings of Turndown
- Computer Models of Turndown
- Turndowns in the Cycles of History
- Emergy Evaluation of Limits in History

The next two are about sustainable society and its resource basis:

- Sustainable Steady State Required by Resource Limits
- Economy Sustained by Substituting Solar Technology for Fuels

The next three concern major change, without turndowns or resource limits:

- Warnings of Major Transition
- Concern About Excess Capitalism
- Predictions of Continued Expansion and Growth Ahead

The last group discusses a smaller economy:

- Suggestions for a Lower-Energy Society

Later chapters in this book use the causal principles in Part II to explain and discriminate among these views.

WARNINGS OF TURNDOWN

In the 1950s, M. King Hubbert (1969) extrapolated the growth of discovered reserves of oil and fuel use, predicting the downturn in oil availability that occurred after 1970. He noted the nuclear wastes, the limited reserves of nuclear fuels (including lithium necessary for fusion), and "not much promise" for solar technology. He wrote, "The rapid population and industrial growth that has prevailed during the last few centuries represents only a brief transitional episode between two very much longer periods . . . of nongrowth."

In 1968, Paul Ehrlich's *The Population Bomb,* a bestseller for public-minded supporters of "Earth Day," warned of the economic turndown, earth damage, and apocalypse caused by overpopulation. In *The End of Affluence* he responded to the oil shortage and inflation of 1974 by reporting indices of danger analogous to miner's canaries and suggested ways for citizens to influence public policy and prepare as individuals. His recommendation was to "change your lifestyle now and avoid the rush later."

In *Awakening from the American Dream* (1976), Rufus Miles explains that energy is the "taproot for affluence"; that a "high energy open society" is "vulnerable to social deterioration, sabotage, and breakdown"; that energy dependence is the "Achilles heel of American society" and excess energy an "unhealthy . . . addiction"; and that only moderate energy use is "consistent with human and ecological values." The ultimate limits to growth, according to Miles, are in "energy resources, technology, and balance of the biosphere." He also states that "further interdependence is unmanageable," that the notion of "one world [is] not achievable or desirable" but "stultifying," and that "economics can't provide policy answers." He expects upheaval of "ideology, ethics, and beliefs" when people realize that the "world's growth policy" is "unattainable."

With artistic images and paradoxes of contrast, W. I. Thompson's *Evil and World Order* (1976) uses erudite humanism to seek the essence of the ascendancy and destruction of the "enlightenment of civilizations." He wrote of the momentary light of the Middle Ages Celtic

Lindisfarne renaissance in Britain as "a catalytic enzyme to effect a change in the entire organism." About the energy basis of civilizations he wrote: "expansion and contraction, expression and compression, expression and suppression . . . from order to disorder . . . the interface between opposites [is] the place of transformation." For our future he wrote that we should "dance out the death of industrial civilization (the iron winter) and experience, not its painful apocalyptic destruction, but its joyous millennial destructuring" and "humanize technology through animism." This vision included the "decentralization of cities, miniaturization of technology, and planetization of nations." He criticized attempts to understand civilization with systems models as "evil," lest somebody try to set up rigid management.

In 1977 N. Georgescu-Roegen suggested that the present civilization could not be sustained because the necessary concentrations of minerals such as metals were being dispersed without sources available to replace losses.

In *Man, Energy, and Society* (1976b), Earl Cook reviewed the state and history of energy resources and "the impossibility . . . [and] undesirability of sustaining the growth state." Cook predicted a three-phase "retrogression to a low energy state" to start in twenty to thirty years. First would be a conservation phase with increased efficiency of transportation and buildings with elimination of waste and luxury. In the second phase there would be a "partial dismembering of the industrial plant," a reduction of services, television becoming "the complete surrogate," methyl alcohol fuel from coal, and widespread use of drugs and alcohol. Characteristics of the third phase would include "most workers back on the farm," a declining birth rate, "social security economically impossible," a "catastrophic population decline [in] Europe," and "world government . . . the most likely alternative." His concerns are that a "hidden energy crisis is one of attitude and institutions . . . shortage of planning and decision capability," and that the struggle for scarce resources could lead to an "upstairs-downstairs" society.

With *The Decadence of the Modern World* (1978), J. R. Sinai wrote of society "sliding into pathology and decadence and at the end might sink into the abyss of a new dark age. . . . A period of decline . . . at a time when leadership ideologies and institutions are failing." He quotes Parson's statement that "the social system is immortal with will to power." He uses psychological terms for pathology, applying them to the larger scale of society.

In *The Lean Years: Politics in the Age of Scarcity* (1980), Richard Barnett characterized transition to a "Post-petroleum World" with predictions of a "global factory by multinational corporations," international military intervention directed to "maintain control over resources," "labor obsolescence[, and] looming world employment crisis of horrendous proportion." His solutions "emphasized resource conservation and stewardship" in addition to "new forms of popular participation."

Citing Easter Island, the Irish Potato Famine, and ecological-population oscillations as models, W. R. Catton's *Overshoot* (1982) explains how "global consumptive growth" will exceed what can be supported. Changing attitudes of "cargoists"–those who assume that a solution to shortages will always appear–could moderate the inevitable crash. He named "Cargoism" after the attitude of native Melanesian islanders who viewed cargo arriving regularly from ships as their magic birthright.

In *The World After Oil: The Shifting Axiom of Power and Wealth* (1983), Bruce Nussbaum describes two alternative paths for turndown: a "dark" scenario with catastrophic disruption and a "light" scenario with harmonious reorganization.

In *The Climax of Capitalism: The U.S. Economy in the 21st Century* (1990), Tom Kemp traces the economy since the Civil War and finds an end to "American hegemony" and a peak in the U.S. economy. He cites such mechanisms as takeovers, increasing debt, and global competition.

In *Beyond Oil: A Threat to Food and Fuel in the Coming Decades* (1991), J. Gever, R. Kaufmann, D. Skole, and C. Vorosmarty analyze diminishing returns of energy-assisted agriculture, energy efficiency, and the declining availability of cheap fuels. They predict an end of economic growth in ten to fifteen years.

In a 1994 article, "The Coming Anarchy," and *The Ends of the Earth* (1996), R. Kaplan describes the apocalypse now spreading in Africa and the disorganizing forces threatening the center of organization in China and Asia as an emerging global crisis. "Overpopulation, scarcity, crime, tribalism, uncontrollable diseases[, and] road warriors" [Somalia] are "global influences weakening nation states [and] destroying the social fabric of our planet." He quotes Homer-Dixon, stating that the Earth is "on the threshold of environmental change as a cause of acute conflict" with "migration of people" caused by "water shortage, [and] depletion of forest, cropland and fish."

In a 1997 review of global oil production and reserves, C. J. Campbell predicts higher oil prices that would reduce consumption by the year 2000, and shortages in Middle Eastern oil developing by 2015, and peak of gas production by 2020. Campbell expects a turndown in the global economy.

B. Fleay's *The Decline of the Age of Oil, Petrol Politics: Australia's Road Ahead* (1995) and C. J. Campbell's *The Coming Oil Crisis* (1997) review the global history of the oil industry, the reserves remaining, and its international politics–vivid and realistic insights on the descent ahead. Walter Youngquist in 1997 reviewed the history and availability of resources required for civilizations, also concluding that the "present trends in population and economic growth are not sustainable."

In *The Coming Age of Scarcity: Preventing Mass Death and Genocide in the Twenty-first Century* (1998), edited by M. N. Dobkowski and I. Wallimann, several authors present the impact and catastrophic possibilities of shortages. One section has case histories of the horrors of disturbed societies in this century in Rwanda, Bosnia, Cambodia, and holocaust Europe. Several chapters explain how a "vicious circle" of development and capital accumulation is approaching a global crisis that will force change. John Gowdy concludes that "it is clear that human activity is beyond an ecologically supportable level. We need to have a declining state before we reach sustainability." Leon Rappoport asks if "it portend[s] a new dark age in which a privileged postmodern minority become smugly detached from chaotic genocidal events afflicting everyone else? Or does it portend . . . a new enlightenment in which a privileged postmodern minority will . . . save what can be saved while expanding the world culture base for a liberating multiplicity?"

In 1999 Jay Hanson arranged a web page (www.dieoff.org) containing quotations, references, and opinions about pending decline.

COMPUTER MODELS OF TURNDOWN

In *Environment, Power and Society* (1971), H. T. Odum uses simple overview mini-models of resources and society to anticipate transition from growth to a society based on limited nonrenewable resources.

In 1972 Donella and Dennis Meadows and associates computer-simulated the world economy reported in the bestselling *The Limits to Growth*. Their graphs of the future crested and decreased around year 2100. They proposed an orderly transition to a "global equilibrium" by restricting further growth and adapting with increases of efficiency and technology. They warned of the potential disaster of further exponential growth and especially population overshoot. The coefficients of the equations of their world model were empirically determined from data correlations before the 1973 energy crisis, and thus they contained the hidden subsidy of cheap energy. When simulated again for a later book, *Beyond the Limits,* in 1992 when coefficients were affected by higher energy costs, the crest appeared early in the twenty-first century.

In 1976 U.S. Congressional Hearings of the Subcommittee on Energy and Power assembled scientists to report on their models of society and energy resources, which were reported as *Middle and Long-term Energy Policies and Alternatives* (94th Congress). Jay Forrester simulated a world model with decline of natural resources and rising population. He showed several scenarios for turndown and quasi-steady state in the next century. Transitions were accompanied by pollution, severe social stress, unemployment, and food shortages. Forrester simulated the industrial sector of his national model and was able to generate cycles of three frequencies suggestive of the observed data (5.5-year business cycle, 16.6-year Kuznets cycle, and 50-year Kondratieff cycle). This model had storage accumulations of inventory, capital equipment, labor, and consumer durables. (The periods of time for internal oscillations are a function of the lags and accumulation times of the storages within a system.)

In the 1976 hearings, H. T. Odum, C. Kylstra, J. Alexander, Neil Sipe, and others simulated a simplistic mini-model of global assets as related to renewable and nonrenewable resources. A peak was found in the 1990s and a secondary peak in 2070. Prices of resources and products generated by that model were used to drive a mini-model of the United States with assets sustained through the twenty-first century.

Also in this volume, K.E.F. Watt, Y. L. Hunger, J. E. Flory, P. J. Hunter, and N. J. Mosman simulated a model of society containing land, agriculture, urban development, and transportation, finding benefit from the effect of higher fuel prices, which limited growth.

In *Groping in the Dark: The First Decade of Global Modelling* (1982), Donella Meadows, D. J. Richardson, and G. Bruckman summarize an international workshop comparing seven large global models. These models and scenarios have built-in designs to reduce limiting effects of resources and achieve goals of growth and development. Some show leveling or turndown; others show continued growth.

In *Environmental Accounting, Emergy and Decision Making* (1996), H. T. Odum includes a simple simulation mini-model, which turns down as part of the pulsing regime that goes with use and regeneration of slowly renewable resources.

TURNDOWNS IN THE CYCLES OF HISTORY

Historians writing on the rise and fall of civilization cite evidences of causality from the intricate details of human history. Their language often implied that descent is bad. Extensively studied were the repeating oscillations observed in business. The rapid oscillations with cyclic periods of a few years make it difficult to recognize the longer-term trends and cycles. Some authors warned of downturns ahead in our society as part of long-range cycles.

S. Kuznets in 1930 and J. A. Shumpeter in 1939 found economic cycles with 3–4, 10, and 50+ year periods. A. G. Frank and B. K. Gillis's *The World System: Five Hundred Years or Five Thousand* (1993) found evidence of longer cycles in five thousand years of human history.

Summarizing his classical compendium of history, Arnold Toynbee's *A Study of History, Abridgement of Volumes I–VI* (1946) found internal societal mechanisms responsible for the ebb and flow of societies. In a chapter titled "The Breakdown of Civilization" he wrote that "many cases of civilization decline . . . can be summed up in three points: a failure of creative power in the minority, an answering withdrawal of mimesis (collective imitation) on the part of the majority, and a consequent loss of social unity in the society as a whole." The decline of civilizations was described as "disintegrations" except where they were incorporated in the rise of another civilization. He wrote that "standardization is the mark of a disintegrating society." Disintegration is by "alternation of routs and rallies." "Petrification," or rigid unchanging society, was given as an alternative to disintegration.

Elsewhere Toynbee described ascendancy due to innovations in organization, productivity, or military efficiency with temporary dominance until the knowledge of the innovations spreads making societies once again equal. After recognizing the cyclic nature of twenty-five previous civilizations, Toynbee hoped that the current western civilizations could be immune to descent and "are not compelled to submit the riddle of our fate to the blind arbitrament of statistics."

With analogous references to ecological systems Paul Collinvaux's *The Fates of Nations: A Biological Theory of History* (1980) relates the rise and fall of civilizations to the warring competition for land resources as the basis for production. The invasion by Genghis Khan, for example, is considered analogous to fire spreading in a grassland. "Aggressive wars are launched by rich societies . . . not of the poor."

After citing Toynbee, D. Elgin's *Voluntary Simplicity* (1981) elaborates on the cycles of civilizations, defining societal characteristics and internal causality in four stages: 1) growth, 2) blossoming, 3) decline, and 4) crisis of transition. He suggested that revitalization can occur in the last stage from the groundswell of people seeking a simpler life followed by growth again. See Elgin's suggestions that follow in Suggestions for a Lower-Energy Society, pp. 55–56.

In *The Downwave: Surviving the Second Great Depression* (1983), Robert Beckman anticipates economic turndown and explains business strategies for profiting during a downwave. Beckman bases his predictions on historical cycles and the similarity of current economic indices to those preceding past turndowns. He derives his recommendations from intricate details of the British economy and by analyses of events in the depression of 1929–1939. The business practice of Andrew Carnegie was cited to show how to be a "winner of the downwave": hold assets to obtain cash before a turndown and use it to buy assets at the bottom of the depression.

In *The Law of Civilization and Decay: An Essay on History* (1986), Brooks Adams proposes a causal sequence: "Velocity of the social movement of any community is proportionate to its energy and mass, and its centralization is proportionate to its velocity." According to Adams, "sooner or later [civilization] reaches the limit of its martial energy, when it must enter on the phase of economic competition," and as a result "invariably . . . dissipate[s] the energy assumed by war." He writes that "capital is autocratic and energy vents itself," it "generates usurers and peasants" leading to a "stationary period followed by disintegration."

C. S. Holling in 1986 and 1995 extended to human societies his ecological cycle concept of "resilience," with its four stages: "periods of exponential change, of growing stasis and brittleness, of readjustment or collapse, and of reorganization for renewal." He described nature as "nested cycles organized by fundamentally discontinuous events and processes."

In *The Collapse of Complex Societies* (1988), J. A. Tainter reviews the history of society and derives a theory that civilizations are "fragile, impermanent things," and collapses are unavoidable. By his concept, growth of complexity (organization, information exchange, massive architecture, arts, trade, hierarchy, military forces, larger integrated areas) aids development of civilization and achievements at first, but later on complexity yields diminishing returns and "unrelenting costs." Collapse occurs "because society invests ever more in a strategy that yields preponderantly less." He said that "economic undevelopment," in order to "live in balance," will not work because the "competition of peer polity" groups would displace those using less.

In *Booms and Busts in Modern Societies* (1990), Koh Ai Tee reviews the economic theories for business cycles and market crashes, considering especially measures for a small country (Singapore) embedded in a global economy. After comparing oscillations in GNP in different countries and their major international economic exchanges, the author concludes that major collapses and depressions are likely to re-occur.

Quoting Ecclesiastes' "to everything there is a season," W. Strauss and N. Howe's *The Fourth Turning: An American Prophecy* (1997) proposes a word model for the cycles of civilization: after a development stage there is an "awakening" to its "high climax." After a stage of "unraveling" is a "crisis climax," when society recognizes its condition.

The transition when society turns at each climax is catalyzed by some unexpected events that galvanize attention. Relating his model to civilizations of history, he found for the United States an "American high" developing from 1946 to 1964, "consciousness revolution" and an "awakening" between 1964 and 1984, and "culture wars" and a period of "unraveling" from 1985 to 2000. He predicts a "fourth turning," the "consciousness of crisis," ahead in 2005. Possible triggers he lists are, among others, a fiscal crisis, a viral epidemic, an event of global anarchy, or a war.

In *The Great Population Spike and After: Reflections on the 21st Century* (1998), W. W. Rostow uses detailed demography including that of China to predict the course of global population surge. Limits to growth inherent in economic structure were considered as suggested by the Kondratief long cycle. After considering an analogous period in the decline of the British economy, he expresses concerns that allowing decline in population growth would lead to economic decline. Global welfare was sought by having the United States "play the role of critical margin in the world's affairs"—with "chaos" the alternative. Strength for this role "depends on our society finding ways to deal with the urban problem." Under the heading "Fusion Power and Its Alternatives," he writes that "resource problems could be much eased if we had a cheap, inexhaustible, nonpolluting source of electricity. This should be a global priority in the twenty-first century."

In *Reorient: Global Economy in the Asian Age* (1998), A. G. Frank recognizes the present major role of the global economy in controlling the hierarchy of nations. He proposes that global economics has been the major factor in all recorded history, previously overlooked by analyses that concentrated on one area of the world and one period of time. His larger-scale theory is a top-down view of nations, using words like globalism, globalogical perspective, and macrohistory. His analysis of current changes in Asia finds the predominance of the "West" to be temporary. The one-world view was described as "Unity in Diversity" and "Diversity in Unity."

In *Guns, Germs, and Steel: The Fates of Human Societies* (1998), Jared Diamond writes of human history as subject to scientific explanation based on large–scale influences. In a global review he relates differences in the rise, sustainability, and displacement of civilizations "not to innate differences in the people themselves but to the differences in their environment." For example, the "fertile crescent and eastern Mediterranean Societies . . . committed suicide by destroying their own resource base . . . an ecologically fragile environment." He writes of other causal influences: the trajectories of military technology, diseases, domestic animals, and decentralized organization.

EMERGY EVALUATION OF LIMITS IN HISTORY

By using *emergy,* quantitative evaluations were made of the resource constraints on past history. The emergy method placed the inputs on a common basis showing what was important.

In "Forest Emergy Basis for Swedish Power in the 17th Century" (1994), U. Sundberg, J. Lindegren, H. T. Odum, and S. J. Doherty find the basis for Sweden's dominance of northern Europe in the seventeenth century in forest resources and iron ore used to process charcoal and steel to develop naval ships, refine silver, and pay for foreign purchases.

Bob Woithe's "Emergy Evaluation of the United States Civil War" (1994) puts native resources, foreign trade, slave populations, economic investment, armament, and the factors affecting the battle at Gettysburg in quantitative perspective. More emergy was eventually available to the north to conclude the war.

Emergy evaluation by H. T. Odum in *Environmental Accounting, Emergy and Decision Making* (1996) shows the international economy stripping the last of the resources on which the economic growth of the last two centuries has been based. He concludes that growth is based on previous resource accumulations now coming to an end.

SUSTAINABLE STEADY STATE REQUIRED BY RESOURCE LIMITS

Especially following the 1973 oil embargo, the mainstream of futurist writing considered growth limitation, the consequence of change, and more efficient society, but most of these did not consider descent. For example, Gray and Martin in 1975 used selected quotes and charming cartoons to condense the essence of sixty-some books and articles that concern a steady state.

In *A Strategy for the Future* (1974), E. Laszlo proposes a "quasistationary global homeostasis, partially changing, never fully attained," a "variable yet enduring goal of human striving." With a "systems view" he involves "law of persistent force" from Herbert Spencer, a structure of independently stable subsystems from H. A. Simon, and social Darwinism to explain the "evolution of complexity, order, self stabilization, self organization, and hierarchilization."

Marvin Harris in *Cannibals and Kings* (1977) uses anthropological history to show how human societies were sustained through cultural adaptation to the particular resources of their area. He traces the origins of society's present characteristics from the interaction between technological innovation, natural resources, and human population densities. Agriculture replaced hunting as population increased and wildlife numbers were reduced. The hierarchy of organized states emerged many times from tribal life when agriculture concentrated wealth, sometimes independently (a "pristine state"). War and infanticide emerged as means of regulating population in relation to landscape. Human sacrifice of captives at an extreme among Aztecs was a mechanism of population regulation. Cannibalism was functional as food and ritualistic enrichment for the upper social hierarchy, where domestic animals were absent. Elsewhere the different cultural attitudes on diet such as use of pork, beef, grains, lamb, milk, fowl, and vegetables were traced to needs for protecting livestocks in special climate regimes such as droughts. Male supremacy and sexual attitudes were traced to a pattern of organization around wars. Capitalism and its overuse of environmental resources started with trade when agricultural produce from one area was sold outside for profit. All these patterns of human society, its learned culture, and psychological training self-organized to fit with special environments. Illustrated by civilizations on rivers is the "hydraulic trap" principle that patterns were frozen when support was stable. Our current civilization was described as an industrial bubble.

William Ophuls's *Ecology and the Politics of Scarcity* (1977) considers what future scenarios are consistent with sustainable ecosystems, energy sources, and the political responses to complexity and scarcity. His verdict on three scenarios were

- [The] high throughput steady state [is] a Faustian bargain fraught with dire political consequences.

- The steady state [in] gratifying as far as possible the materialistic and hedonistic appetites of the populace contravenes the lessons . . . of political philosophy . . . and ethical spiritual teachings.

- [The] minimal frugal steady state . . . would lead toward a decentralized Jeffersonian polity of relatively small, intimate, locally autonomous, and self-governing communities rooted in the land or other local ecological resources.

Ophuls and Boyan restated sustainable ideals in 1992.

Stephen Lyons's *Sun: A Handbook for the Solar Decade* (1978) contains a broad range of viewpoints from its twenty-six authors on the level of civilization desirable or possible on solar energy. Ivan Illich proposes "the choice of a minimum energy economy," stating that "above the threshold, energy grows at the expense of equity" and calling for "progress at the speed of a bicycle." Murray Bookchin wrote to reintroduce nature into a "technology for life" and of a post-scarcity society in which humans can have "free relations to nature." Other authors sought to substitute more energy from biomass production and solar technology to sustain civilization.

In his 1978 thesis "In Search of Steady State," Ivan M. Johnstone used energy systems concepts and details on resources and growth to consider what a steady state could be for New Zealand, with the "highest potential carrying capacity surplus ratios in the world." He found carrying capacity to be dependent on the "consumer level of life" adopted and the "optimum organizational size of sub-settlements."

Global 2000 Report to the President, a detailed four-volume comprehensive review of world resources and their effect on the future, was conducted in 1980 by a Carter Administration task force led by G. O. Barney. An appraisal of some of the complex world simulation models was included. Because the extrapolated growth of populations was greater than the carrying capacity of the Earth's resources, economic slowdown was expected.

Hazel Henderson's *The Politics of the Solar Age: Alternatives to Economics* (1981) characterizes the transition to a solar age of "wood, water, wind, sun, oceans, [and] biomass" as "trial by entropy." She predicts a "coming era of post-economic policy making" with "leaders and decision makers no longer in charge," a "repeal of the divine right of capital," a "revolution from hardware to software," and emphasis on the "rights of the child."

In *Envisioning a Sustainable Society: Learning Our Way Out* (1989), Lester Milbrath compares the beliefs and values of a "failing" dominant social paradigm (DSP) with a new environmental paradigm (NEP) offered as sustainable. To accomplish transition he proposes for each nation a "new institution: the council for long-range societal guidance," the purpose of which is to aid in our "learning our way to a new society" and a "consciously anticipatory . . . transforming [of] the dominator society." It should include "ecojustice" (*ecological wholeness* and *social justice*), a "health structure that emphasizes wellness," "electronic town meetings," "soft energy," "cogeneration," a "planning structure" that "avoids large projects," a "production structure that emphasizes sustainability," and "a science court" that would make rulings on scientific accuracy.

Thomas Hine's *Facing Tomorrow: What the Future Has Been, What the Future Can Be* (1991) seeks "subtle progress" to shape technology to advance human aims rather than reshaping humans to fit technology. Hine acknowledges that "today people become angry at the future because it is not going to provide what was once expected." He seeks a synthesis of innovation and basic values so that we "can once again feel the exaltation of moving toward something we want."

In *Ecological Economics: The Science and Management of Sustainability* (1991), Robert Costanza introduces thirty-two papers with a statement of goals and policies for a sustainable economy: introduce ethical support, maintain natural capital, link revenues and uses, and de-bureau-cratize institutions.

In *Paradigms in Progress: Life Beyond Economics* (1991) and *Building a Win-Win World: Life Beyond Global Economic Warfare* (1996), Hazel Henderson scans hundreds of expressions and events to read the pulse of society beginning to turn its emphasis away from competition for money. As a replacement, she sets out what is sustainable. She writes of "cultural . . . codes and biodiversity as the real wealth of nations" and "information [as] the world's real currency." Determining cultural codes, analogous to genetic DNA codes, is needed. To better under-stand the information revolution confused by "infoglut, government by mediocrity, and attention economy," she classifies information on five scales of decreasing quantity and increased quality: (a) raw data and unpatterned facts, (b) assumption models, (c) worldview concepts, (d) goals and purposes, (e) values.

Henderson criticized recent indices of "wealth and progress" in-vented to replace monetary indicators including:

- A World Bank *New System of National Accounts* that included (a) natural capital, (b) produced assets, (c) human resources, and (d) social capital.
- A United Nations Development Programme (UNDP) *Human Development Index* (HDI) that combined (a) life expectancy, (b) educational attainment, and (c) basic purchasing power.
- A UN Statistical Division index: *Environmentally Adjusted Net Domestic Product* (EDP).
- H. Daley and C. Cobb, *Index of Sustainable Economic Welfare* (ISEW) (Daley 1996).

Combinations of economic and environmental measures "should not be aggregated . . . a tortured definition . . . arcane assumptions behind a one-number index." Instead she offers a set of indicators, the *Country Futures Indicators* (CFI) with nine indices of system sectors and four-teen indices of achievement of social goals. She suggests that "the new scorecards[, by] raising the economy's ethical floor[, can generate] deeper politics of meaning."

After detailing the dysfunction and global danger of our excess consumerism, A. Durning in *How Much Is Enough?* (1992) calls for a "culture of permanence" with the ecological equivalent of the golden rule–that "each generation should meet its needs without jeopardiz-ing the prospects for future generations."

In *Living within Limits: Ecology, Economics and Population Taboos* (1993), Garrett Hardin assembles principles, historical roots, and examples, ancient and recent, to understand resource limits. He warns of the excess population during the "demographic transition," in which death rates decline faster than births, and the consequences of "population crash." He calls for "a national demostat"—a self-regulating population-control mechanism. He reviews the history of radical means for population control: isolate overpopulated nations from global food aid; impede emigration from overpopulated areas; allow infant mortality, infanticide, and abortion.

With *The Ecology of Commerce: A Declaration of Sustainability* (1994), Paul Hawken (quoting Henry Wallace) suggests "interdependence" as the way to make business and environment symbiotic. He wrote that

> the global economy has already exceeded carrying capacity—that point beyond which further growth will decay and effectively destroy its host. . . . Business must change its perspective and its propaganda . . . the counter myth of no limits.

He suggests "green taxes" on environmental uses such as salmon harvest, and tax money to help sustain environmental functions.

In "Ecological Footprints and Appropriated Carrying Capacity: Measuring the Natural Capital Requirements of the Human Economy" (1994), W. Rees and M. Wackernagel use the concept of necessary support area, the "ecological footprint," to dramatize the limited carrying capacity of the Earth for civilization. "Far from growing with the expansion of the urban world," they write, "the resource base sustaining the human population is in steady decline." They write that "the appropriation of most of the world's carrying capacity by the urban industrial North" and "the insistence by the South of its right to a fair share (and the threat to seize what it can through sheer growth) . . . were the only issues at the Earth Summit in Rio in June 1992."

Edited by R. Krishnan, J. M. Harris, and N. R. Goodwin, *A Survey of Ecological Economics* (1995), an anthology of 102 selected writings, contained widely varying perspectives on environmental limits and their relation to the economy.

ECONOMY SUSTAINED
BY SUBSTITUTING SOLAR TECHNOLOGY FOR FUELS

The dilute nature of solar energy makes the concentrating process required for its direct use inherently costly. However, many futurists based their recommendations on the assumption that solar energy can substitute for the present level of energy use now running on fossil fuels. (See Chapter 9.)

In *Energy: The New Era* (1974), David Freeman analyzes energy sources, international energy policies, and energy pricing. Energy conservation, limitation to affluence, increased durability of products, and increased efficiency were indicated for the period beyond 1985. He

"GROWTH FOR ITS OWN SAKE — SLOGAN OF THE CANCER CELL" — EDWARD ABBEY

Courtesy Joel Pett. Reprinted with permission.

suggests that "perhaps the ultimate step is to harness the sun directly or to control on Earth the fusion reaction that occurs in the sun."

With *Soft Energy Pathways* (1977), A. B. Lovins proposed that "soft" energy sources, solar and wind technology, conservation measures, and lower energy living could fully replace "hard sources," nuclear energy and fossil fuels.

In *The Politics of Energy* (1979), Barry Commoner describes limits to nuclear energy, but assumed solar technology had enough net energy to run a nation. His first law of ecology is that "everything is connected to everything else."

Robert Stobaugh and Daniel Yergin edited *Energy Future: Report of the Energy Project of the Harvard Business School* (1979), an analysis of energy alternatives and policy. Noting the "end of easy oil," they sought "a balanced energy program . . . to replace overdependence on imported oil" and "energy wars." They proposed that "conservation [is] the key energy source." The "promising but still largely untested domain of solar [power]"–power towers, solar satellites, solar heating, solar voltaics–was predicted to be "10–30 percent of national energy by year 2000." They stated that "serious effort should be made . . . to

sustain the nuclear option." The "long term focus" should be "to utilize coal more cleanly either in direct combustion or as a liquid or
gas." Cost curves for "increasing difficulty of exploitation" need to be
incorporated in previously overoptimistic national energy models.

A popular "solar lobby" supported solar technology as clean and
efficient, but in *Energy Risk Assessment* (1982), Inhaber assembled extensive quantitative evidence that solar technology indirectly impacts
society, the environment, and the economy more than fossil and nuclear
energy.

Lester Brown and associates with the Worldwatch Institute have
issued special topic books and position papers since 1974 and yearly
State of the World books starting in 1984. Evaluations and projections of
the state of world resources were followed with interpretations regarding a sustainable economy's uses of land area, soil, water, fertilizer,
forest and fishery products, wastes, conservation, recycling, and miniaturization. For example, they wrote that "ecological deficits cause
economic deficits" as soils and woods of the world are used faster than
they are generated. They assumed that technology of renewable solar
and wind energies could sustain developed society.

Power Surge: Guide to the Coming Energy Revolution (1994), a
Worldwatch Alert Book by C. Flavin and N. Lenssen, assembled ar-
guments that solar technology and wind can become the major energy source for the modern society. (See Chapter 9.)

In *State of the World 1999, Millennium Edition* (Worldwatch, 1999),
Lester Brown and associates suggest an economic boom can come if
government and private enterprises invest in recycling and renewable
energy to head off the disasters of environment and resource depletion.

J. F. Coates and J. Jarret, editors of thirteen chapters in *The Future:
Trends into the Twenty-first Century* (1996), write of their "belief in society's
ability to explore and take responsibility for influencing the future."
Writing on the future of the environment at risk due to overpopulation, Richard Lamm wrote that it is "hard to motivate people when
there is uncertainty . . . when the problem will manifest itself in the
future." Writing of governance, J. M. Grill and Gary Cappert were
concerned for "polarization of society into single focus constituencies."

WARNINGS OF MAJOR TRANSITION

Many public affairs books tracked the emergence of our complex information society and assumed resources would continue to be available. But because of serious problems with the present systems, the
authors proposed or predicted major changes and new patterns.

Written in a time of expanding atomic energy, W. F. Cottrell's
Energy and Society (1955) considered changes in social values in going

from a "low to high energy society" and strains in the economy from unequal energy access. He wrote that "energy limits what man can do and influences what he will do."

Daniel Bell, in *The Coming of Post-Industrial Society* (1973), writes of "an end to scarcity" but a "shift to no growth," an "information society." He expects society to turn from dependence on "industrial and natural order" toward "a game between persons"; predicts "a shift from manufacturing to service" but sees service as "a drag on productivity," and expects "a new science-based technology" without energy limits and a "rise of a technological elite."

Fritjof Capra's *The Turning Point: Science, Society, and the Rising Culture* (1982) relates the major global transitions in progress to cyclic theories of society. He quotes the *I Ching* (Wilhelm, 1968): "After a time of decay comes the turning point . . . movement is natural, arising spontaneously . . . the new is introduced." He used Pitirim Sorokin's theory for culture's "waxing and waning": "cyclical rhythms of interplay between [the] sensate" stage, which senses reality, and the "ideational" stage, a period of internal imaging, "produce an intermediate, synthesizing period," an "idealistic" stage. Capra observed three major areas of the global transition ahead: 1) "disintegration of patriarchy," or male dominance, in favor of a balance of "yin and yang," 2) "transition from the fossil fuel age to a solar age," and 3) "paradigm shift" from "mechanistic and reductionistic . . . scientism" to a "systems view [of the] interrelatedness and interdependence of all phenomena, physical, biological, psychological, social, and cultural," that "transcends current disciplinary and conceptual boundaries."

In *The Synergism Hypothesis: A Theory of Progressive Evolution* (1983), Peter Corning explains that "synergistic effects . . . constitute the underlying cause of the directional aspect of evolutionary history" and the "progressive emergence of complex, hierarchically organized systems." He writes that "political regression–the simplification, dismemberment or collapse of cybernetic social processes–is always associated with a decline or loss of functional synergism." He concludes that "we need more politics, not less of it."

In *The Eighth Day: Social Evolution as the Self Organization of Energy* (1988), Richard N. Adams found "complexity, hierarchy, and energy dissipation" to be characteristics of many cultures. "Triggers evolve to enhance and promote a continuing expansion of human societies." "Humanism is one (trigger) today," Adams states, and a "viable society keeps the energy cost of triggers low."

Paul Kennedy's *Preparing for the Twenty First Century* (1993) uses a "human development index" combining "life expectancy, literacy rate, and per capita GNP." On this index the United States rated 112, Japan

130, and Sweden and Switzerland 128. He describes the tension between nation-state and economics. He finds three key elements to prepare: "The role of education . . . the place of women . . . and the need for political leadership."

J. Madrick in *The End to Affluence* (1995) tracked the favorable conditions for rapid growth and affluence in the history of the United States's economy. But he found growth rate and affluence since 1970 being reduced by international competition, mismatch of education and jobs, fragmentation of industries, debt, inadequate capital investment, and political and public unwillingness to face realities.

In *The End of Work: The Decline of the Global Labor Force and the Dawn of the Post Market Era* (1995), Jeremy Rifkin found that the number of people exceed the number of available jobs. He also described a "Third Society" and proposed to tax the information business for the money to support people through "social communes." He saw "government increasing its role with the Third Society with less role in business."

In *The Clash of Civilizations and Remaking of the World Order* (1996), Samuel Huntington finds "modernity" developing all over the contemporary world causing clash and struggle among nine main societies (Western, Latin-American, African, Islamic, Sinic, Hindu, Orthodox, Buddhist, and Japanese). He suggests that world populations are being recombined in these new "civilizations" according to their religious, geographic, and cultural histories. He believes world order and peace will come from "reaffirming [the] identity of western civilization" while "cooperating to maintain the multicivilizational character of global politics." He writes of the difficulties and fear of any unified "world civilization . . . yielding to barbarism, generating the image of an unprecedented phenomenon, a global dark ages possibly descending on humanity."

With *Turning Away from Technology* (1997), Stephanie Mills reports on two "megatechnology" conferences that advocated less technology as a vision for the twenty-first century. Mills criticizes the "pathos in frivolous gadgetry" and "maldistribution of appropriate technologies," and states her beliefs that "megatechnology undermines and overturns authentic, more equitable existence" and that "technology will not solve problems caused by technology (technofix)." She cites "development depredations" that include "attempts to privatize what remains of the commons . . . wildlands, crop varieties, and megadams." Cited as "technofantasy" are proposals for a "hydrogen economy, genetic engineering, nanotechnology, and autonomous technology." Fear is engendered by the "precariousness of megatechnological infrastructure," the "totalitarianism of trade and technology," and a "global market poised to lift off by technological means leaving denuded fields and hillsides, festering favelas, flooded coastlines, and billions of redundant human beings." The conferences found that solutions lie in "economic diversification and decentralization" and in the "devolutionary work ahead; from the mechanical to the organic," but also found that "understanding of the root causes is lacking."

A. Bell in *The Quickening: Today's Trends, Tomorrow's World* (1997) warns of rapid changes: "Some parts are positive–many others are not." Bell states that "right now the world is getting progressively more chaotic and dangerous," and asks, "do we really think that our children's children will have what we now have? Not likely and I'm not just talking about gas and water." His stated premise was that there is adequate energy for 260 years.

In *New Rules for the New Economy: Ten Radical Strategies for a Connected World* (1998), Kevin Kelley explains the emergence of information networks as a dominant influence of the future not following usual economic principles. He suggests that information supply increases with decreasing price; there is self-reinforcing success by quadratic acceleration (growth proportional to N^2 where N is the number of network nodes). Networks can be compared to ecological systems with units surviving by contributing to network expansion, and innovations prevail that are first distributed free. He expects unlimited takeover by decentralized information networks. A caveat was given: "Of course, all the mouse clicks in the world can't move atoms in real space without tapping real energy, so there are limits to how far the soft will infiltrate the hard."

CONCERN ABOUT EXCESS CAPITALISM

In *The Death of Capital* (1977), Stephen Harris perceives "the last great merger," that of "economic and political systems." This would create a "joint national enterprise" with "government protecting the rights of capital[,] . . . bureaucratic dead weight," "unemployment," and "open conflict on economic policy." Tawney states that "faith" has changed from "people to mechanism" (markets). T. Veblen's notion that "business [is] the enemy of industrial growth" agrees with Shumpeter's statement that "capitalism [has been] killed by its achievements" and Ricardo's belief that "wages and profits [are] depressed with full development." John Stuart Mill, however, states that "stationary capitalism implies no stationary state of human improvement."

Robert Heilbroner's *Beyond Boom and Crash* (1978) refers to B. Horvar's stages of capitalism: competitive capitalism from 1700 to 1875; monopolistic capitalism from 1875 to 1930; and welfare capitalism from 1950 to the present.

In *The Cultural Contradictions of Capitalism* (1976) and a postscript (1998), Daniel Bell described the confused state of the present capitalism-dominated culture as "unstable [and] transitory." Quoting Max Weber's *The Protestant Ethic and the Spirit of Capitalism* (1905), Bell traces capitalism's start from frugal savings and individualism after the Protestant Reformation. But in 1996 Bell found the "Protestant work ethic overwhelmed by acquisitiveness," or consumerism. An early trend of capitalism was to unify culture with a simplifying "modernism: the self-willed effort of a style . . . to remain in the forefront of advancing consciousness." But in 1996 he found uncontrolled complexity he described as "PoMo," or "post-modernism," a reaction in which "anything goes." He wrote that the "crucial sociological issue for the society" is the "separation of law from morality." He cited a critical decision in the United States Supreme Court in the late nineteenth century when a "corporation was declared to be a 'legal person.'" Thus capitalistic economy was given the freedom from control that the Constitution intended for individuals. He described the "unraveling of the middle class," "culture wars," and "a rising distrust of politics and law" as characteristics of the state of the nation.

After finding economic evils in paper entrepreneurism and other free market ideologies that switch money without generating wealth, Robert Reich in *The Next American Frontier* (1983) suggests changes in government structure and programs. With regard to the business cycle he wrote, "economic decline hardens the resistance to change because more people are made more vulnerable to the risks attendant on change. . . . Such feelings engender a widespread conservatism [that says] let us at least preserve what we have, retreat to the ways we used to do things and return to 'basics.'" Reich did not regard decline as inevitable.

F. Fukuyama's *The End of History and the Last Man* (1992) quotes Aristotle, describing an "endless cycle of one regime replaced with another." Fukuyama proposes that society can rise above the pessimism of history with a "coherent and directional history of mankind that will eventually lead the greater part of humanity to liberal democracy" by meeting two primary needs, as stated by Thomas Hobbes and G.W.F. Hegel, respectively: (1) the necessities of life support and liberty, and (2) the need for recognition, a sense of justice, and a share of political power.

In *Vision 2020: Reordering Chaos for Global Survival* (1994), E. Laszlo saw an "end to placid growth—heading for period of chaos." He proposed a "third strategy" between laissez-faire and centralism: retaining the power of nations, but with "a concord of (international) environmental cooperation."

With *21st Century Capitalism* (1994), Robert Heilbroner writing on "economic determinism" defined *wealth* as "symbols of power and prestige," not of virtue. He predicted future trends would be instability and collapse of uncontrolled capitalism that could be "stabilized with Keynesian policies," "public investment," and a "spectrum of capitalism" like Swedish socialism.

In *Jihad vs. McWorld: How Globalization and Tribalism Are Reshaping the World* (1995), Benjamin Barber found everyone becoming homogenized as a world consumer, causing many to seek a regional tribal identity and saw both trends as threatening to democracy. Worldwide spread of McDonald's was used as the symbol of globalization, which includes "teleliterature . . . music television . . . brand choice . . . ideology of entertainment . . . global pop culture . . . malls as entertainment plazas . . . media merger frenzy . . . everything belonging to everything else . . . cultural warfare in the name of markets[, and] boundaries permeable for the good or for the bad." *Jihad* was used as a symbol of the rejection of global influence by local extremism with "microwars . . . Balkanization . . . anarchy . . . civil war . . . privatization . . . theocracy . . . parochialism . . . exclusion and resentment, [and] new world disorder."

Justice of energy indices were calculated for countries to show the international injustice, struggle for power between global and regional trends, and loss of western autarky: JEDI-A = percent of world energy divided by percent of world population; JEDI-B = percent of world energy divided by percent of world economic product.

In his *One World Ready or Not: The Manic Logic of Global Capitalism* (1997), William Greider cites example after example of the destructive behavior of global capitalism and free trade: subversion of national policies by control of debt and bond interest by international capital; creation of worldwide excess of production that consumers with declining income cannot buy; transfer of jobs and investment away from developed countries; leverage on governments to provide tax incentives; amassing of capital and power in the hands of a few; and social and environmental damage. He warns of conditions ripe for a global economic crash comparable to the Great Depression of 1930, which was followed by political upheaval and the rise of dictatorships.

Greider suggests that "global Keynesian economics is needed by which the expenditures of governments consume surpluses, create market demand, promote rising wages, redistribute wealth downward to those who will spend" and use labor to aid industrial ecology and the environment. According to Greider, "nations would have to find the courage to reclaim control over free-running capital." In place of central banks making low interest loans to other banks to increase the money supply during growth, he suggested that "if newly created money was instead lent directly to citizen-ownership trusts, it would provide very cheap capital for a large public purpose." He recommended "tax capital in place of labor," the use of "environmental taxes if they can be equitably applied," use of "government procurement" to force "industrial reform," "forgiv[ing of] the debtors," and a "refocus[ing of] national economic agendas on the priority of work and wages, rather than trade or multinational competitiveness as the defining issue of domestic prosperity." His measures assumed continued economic growth.

W. Wolman and A. Colamosca in *The Judas Economy: The Triumph of Capital and the Betrayal of Work* (1997), G. Soros in *The Crisis in Global Capitalism* (1998), and E. Yergin and J. Stanislaw in *The Commanding Heights: The Battle between Government and the Marketplace* (1998) also express concerns for crisis in the global society because of the excesses of unregulated international capitalism. As Yergin and Stanislaw wrote about the trend of "beliefs and ideas [moving] away from the traditional faith in the state and toward greater credibility for the market."

In *The Ownership Solution: Toward a Shared Capitalism for the 21st Century* (1998), Jeff Gates recommended accelerating a recent trend to spread stock ownership and their earnings throughout the population. He wrote of "history's most enduring danger—concentrated ownership." He suggested mechanisms such as paying workers with stock (employee stock ownership plans, or ESOP), buying stock in utilities as part of payment for use (consumer stock ownership plan, or CSOP), and a plan for those using public resources to pay shares of stock to citizens (a general stock ownership corporation, or GSOC). He suggested that shared ownership is a way to retain democratic control and equity within global capitalism.

PREDICTIONS OF CONTINUED EXPANSION AND GROWTH AHEAD

In "An Operating Manual for Space Ship Earth" (1968), R. Buckminister Fuller laments the waste of fossil fuels, but sees rising carrying capacity because of the increasing "synergism" (organization) and "information use," which makes it possible "to do more with less." He writes that society can "enjoy and explore the universe as we progressively harness ever more of the celestially generated tidal and storm generated wind, water, and electrical power concentrations."

In the books *Future Shock* (1980) and *Creating a New Civilization* (1994), A. Toffler and H. Toffler report accelerating change to a third wave of civilization emerging from the industrial, urban, and economic second wave, which had displaced the rural agricultural first wave. The "third wave" includes an emphasis on information, diversity, complexity, smaller units, electronic family cottages, faster economic dynamics, and the mechanizing and diversity of energy sources.

M. Cetron and T. O'Toole in 1982 made detailed forecasts into the twenty-first century: lengthening life, robotic displacement of blue-collar work, increase of environmental technicians, sun belt expansion, and voracious use of fossil fuels. Expecting fusion, breeder, and other nuclear energy development, turndown was not predicted. The pope was expected to eventually accept birth control.

In 1984, Simon and Kahn edited *The Resourceful Earth: A Response to Global 2000* (Barney 1980), a twenty-one chapter volume citing unused capacity of water, food, forest products, and fishery products. Faith in continued growth was based on continuing nuclear power. Decreasing prices of resources earlier in the century was interpreted as proof that resources were becoming more available. (See Chapter 10.)

John Naisbitt's bestselling *Megatrends* (1985) and *Global Paradox* (1996) documented trends he believed would continue: short-run fads, unsustainable trials, excesses of the affluent, subsidized experiments, dramatization of new devices, chamber-of-commerce advertising, booms of defense spending, Star Wars largesse, robot manufacture of large cars, lawsuit chaos and legal waste following deregulation, fancy and unaffordable technology in homes, mergers towards large-scale chains, more Silicon Valleys, lower per-capita private and public funds for health and retirement, increasing employment of women and use of day care centers, markets that make work more efficient, further spread of inexpensive microcomputers and videotape processing, private profit from high-tech medicine, and sun-belt expansion based on retirement funds and tourism. His stated paradox is that there is global unification while at the same time a strengthening of individuals and small groups.

Summarizing *The Global Possible: Resources Development and the Next Century* (an international workshop proceedings), Robert Repetto in *World Enough and Time: Successful Strategies for Resource Management* (1985) wrote that sustainable development and continued growth were possible through increasing efficiency of resource use, pricing and taxing for conservation, developing management capability, preserving natural capital, and stabilizing population. Many papers in the workshop warned of disappearing resources and unsustainable practices. The reports acknowledged the need to find new energy sources to replace oil within twenty-five years.

Daniel Bell in "Introduction: Reflections on the End of an Age" (1995) writes of "the post industrial society" finding five "dimensions":

• economic sector
• occupational distribution

- axial principle (theory and policy formulation)
- futures orientation–technology and its assessment
- decisionmaking–"new intellectual technology"

He writes that "no society can ignore the problem of balance leaving the basic decisions either entirely to the market or to bureaucratic rule . . . and technology provides no answers."

The authors in *The State of Humanity* (1995, Julian Simon, ed.) write on population, health, standard of living, productivity, poverty, resources, agriculture, pollution, environmental systems, and public attitudes up to 1990. Many graphs showed increased indices of value and benefit during the century. Most but not all showed expansion and growth to the end of the period. Simon summarized the book by writing that "there is stronger reason than ever to believe that these progressive trends will continue past the year 2000, and indefinitely." Pessimistic writings by previous authors were attacked. For example, Simon and Wildavsky discussed published estimates of species loss, concluding that "the highest proven observed rate of extinction until now is only one species per year."

G. Celente in *Trends: How to Prepare for and Profit from Changes of the 21st Century* (1997) sees emergence of a new "Theo-economic faith," in which "economy is good" and the "idea that profit was not the sole arm of corporate existence." He sees no limits, predicting an "energy revolution" based on cold fusion having a net contribution of ten times more energy than is used by the process. (See Chapter 9.)

In *Bold New World: The Essential Road Map to the Twenty-First Century* (1997), William Knoke predicted "the age of everything-everywhere," a "placeless society" with worldwide decentralized organizations: "turbocapitalism . . . globocorps . . . ecotribe[, and] mail order in place of malls." He wrote of the "collapse of hierarchies . . . erosion of governance . . . nation-states [becoming] obsolete" and saw "the coming amoeba . . . global government . . . empower[ing] the individual."

In *Is Progress Speeding Up: Our Multiplying Multitudes of Blessing* (1997), John Templeton wrote of the present as "a wonderful time in history," stating that "pessimism [is] unwarranted but stubborn." He cites "increased employment . . . high life expectancy," and the "first mass upper class." Templeton writes that "50 million Americans now own stock" and are "beneficiaries of upsizing." He also cites food's turn "from scarcity to plenty," and "declining diseases . . . more interesting jobs . . . boon for business[, and] political [and] economic freedom." Education has a "shining record," women are making "great strides," and religions are experiencing a "growth of believers." He described recent "decreases in energy requirements per unit of economic output," "decarbonization" (the decrease of carbon dioxide release per unit energy), and "increased environmental quality." According to Templeton, "the greatest challenge . . . will be to achieve a more equitable distribution." "If a well-established past is any sort of prologue, the technological explosion in information and communications appears to have an almost boundless future."

In *The Great Boom Ahead* (1994) and *The Roaring 2000s: Building the Wealth and Lifestyle in the Greatest Boom in History* (1998), H. S. Dent, Jr., saw unqualified expansion ahead, as did Bob Davis and D. Wessel in *Prosperity: The Coming Twenty-Year Book and What It Means to You* (1998) and Knight Kiplinger in *World Boom Ahead: Why Business and Consumers Will Prosper* (1998).

W. Michael Cox and Richard Alm's *Myths of Rich and Poor: Why We're Better Off Than We Think* (1999) gathers statistics to refute society's "prevailing pessimism." The authors accept as fact that in the United States hourly wages decreased 15 percent since 1973, but they show that the total income per person has risen because more people are working longer. The authors judge increased benefit from the increases of consumer assets per person such as houses, autos, auto accessories, household appliances, medical advances, and food. Many of their graphs of consumer assets level off in the 1990s. The United States was higher than other leading countries in selected indices, including gross domestic product per person, life expectancy, and college-educated population. Downsize firing was justified by showing the increase in high-paying service jobs and how people moved to new industries in the years of rapid growth. After pointing to U.S. leadership in technology, the book concludes that "the best is yet to come."

SUGGESTIONS FOR A LOWER-ENERGY SOCIETY

H. T. Odum in *Environment, Power and Society* (1971) uses the analogy of economies and ecosystems such as Silver Springs, Florida, to explain how complex systems expand and recede without crashing. Means of adapting to less energy suggested trends: "preservation of knowledge and cultural memory, smaller cities, fewer cars, more agricultural workers, less pollution, crash efforts at birth control, ecological engineering, less terrible war, and revised religious doctrines in support of world balance."

Verbal and poetic, Paul Williams's *Das Energi* (1973) identifies values with the flow of energy. The "human place is in the flow of the Earth's energy." For the large scale he writes: "refuse to have anything to do with any thing that seeks to grow wealthier rather than healthier, larger rather than truer."

Ernst F. Schumacher's *Small Is Beautiful: Economics as if People Mattered* (1973) warns of dangerous trends, misplaced goals, and the "metaphysic" of those economic systems that "dominate and conquer nature," "ignore man's dependence on the natural world," and cause "destructive forces . . . terrorism, genocide, breakdown, pollution, exhaustion." He is "sad at [the] loss of classical Christian heritage."

He believes that simply expanding the economy could not solve problems. He writes that society is "unable to achieve ideals using the 'road to riches' " or obtain "peace from prosperity." He writes that "simplicity and non-violence are . . . closely related." Attempts to derive "the good, the true, the beautiful" from the "pursuit of wealth and power" is "crackpot-realism." But he criticizes those who would only "substitute non-growth for growth, that is to say, one emptiness for another."

He states that the "most fateful error of our age" is basing "production" on disappearing "natural capital." He suggests means to "build up soil fertility" to produce "health and permanence." He describes successful "intermediate technology" examples such as small-scale oil refineries in Africa and small manufacturing plants in Puerto Rico. Regarding the meeting of electrical needs, he writes that no one should "put all his eggs in the nuclear basket . . . before he discovers that a solution [to radioactive waste] can be found."

For a solution to "mass migration to cities" he suggests an "agro-industrial culture." In place of "large-scale private enterprise" he suggests a new kind of "cooperative" that could represent the "public hand . . . by means of a fifty percent ownership." It "would receive one-half of the distributed profits" that a "Social Council" would manage.

Since "Man is small, and, therefore small is beautiful," he suggests that a reorganized economy could substitute ideals of a "Buddhist economics[:] maximum well being with the minimum of consumption," involving the "purification of human character" rather than the "multiplication of wants." He sees this as a way that "enhances a man's skill" and develops a lifestyle for which, he concludes, "any organization has to strive continuously[:] orderliness of order and the disorderliness of creative freedom" but with the "guidance [that] can't be found in science or technology." These ideals would accord "secondary . . . place . . . to material things" and herald a "return to meta-economic values" and "peace and permanence."

In *Middle and Long-term Energy Policies and Alternatives* (1976a), Earl Cook documents the crest of known U.S. oil reserves in 1967, the leveling of efficiencies in energy use, and the accelerating costs of energy-environmental interactions. He examines three scenarios for the future: Of Alternative I–Continued High Energy Growth, Cook states that there is "nothing known to science to preclude it" nor "to guarantee it." However, he warns of the dangers to society of "large amounts

of energy with net energy profit." He included excess population, au-
tomation, economic inequality, unemployment, developments that
"exceed capacity of the environment to absorb wastes," scarcity of
environmental recreation areas, the "pervasive ethic of pleasure," and
"threat of revolt." Regarding Alternative II–Transition to an Almost
Steady-State Economy, Cook cites John Stuart Mill's "stationary state"
with changed religion about growth, increase in efficiencies, reduced
wastes, stern measures for population limitation, and social struggles
for limited resources. Cook writes of the possibility that reduced energy resources will cause Alternative III–Retrogression to a Low-
Energy Economy, which would include the "restricting of individual
energy uses . . . extreme efficiency measures . . . [a] shrinking service
sector"; "television [becoming] the complete surrogate" for mass recreation; less-intensive manufacturing; decreased protein diets; people,
including the elderly, back on farms; dropping birth rates; and "no
unemployment."

With *Energy Basis for Man and Nature* in 1976 and 1982, H. T. Odum
and E. C. Odum gave more details on transition and a lower energy
future, recognizing the necessity for pulsing, ecosystems as "solar energy partners," possibility of the "steady state as a happy place," and
thirty-five characteristics of a steady-state economy at a lower energy
level.

In *The Sane Alternative: A Choice of Futures* (1978), James Robertson
calls for a new equilibrium society that is "sane, humane, ecological,"
based on small-scale social relationships, an informal economy, "sup-
portive balance with world ecosystems," and cooperative group choices
to avoid four undesirable scenarios: "business as usual, catastrophic
breakdown, totalitarianism after breakdown, and expansionist high
technology."

W. J. Davis in *The Seventh Year: Industrial Civilization in Transition*
(1979) made predictions and recommendations for transition to an
energy-scarce society: decentralization, less government, less university knowledge and research, more employment, less leisure, more
people in agriculture, less specialization, more general education,
diffused culture, increase of environmental values (which he called
"ecologic"), and less money and wealth. He saw lower energy as an
opportunity that "need not be uninviting."

D. Elgin's *Voluntary Simplicity* (1981) gave the following characteristics of a revitalizing period after decline: consumerist revolution for
more moderate consumption, solar power–renewable energy, energy
conservation, communication in place of travel, efficiency in transport, localized markets and entrepreneurial initiative, migration to open
areas, growth of information-technology-service, local agriculture, more

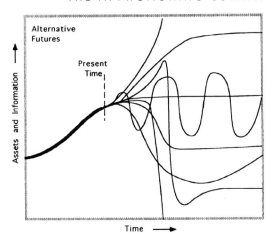

Figure 3.1 Trends of assets and information according to scenarios proposed or predicted in Chapter 3.

recycling, worker-influenced management, nonpolluting engineering, do-it-yourself activities, more preventative medicine, urban villages with food-producing land use, efficient housing, diversity-honoring social order, cross-cultural sharing, decentralization with globalization, redefinition of the good life and its artistic expression, and shift of political perspective from "egoistic-materialistic-nationalistic" to "transpersonal-material-spiritual-planetary." In his Appendix was a model for the limits of knowledge affecting the summit of civilization. Knowledge needed to operate society accelerates with complexity until it exceeds society's ability to absorb and use it. Knowledge has diminishing returns with scale.

With the format of a student's course study guide, E. Jernigan's *America and the World 1995–2015* (1995) suggests a meritocracy, and states that "green is beautiful," that public health is a big investment, that we should "save energy for business" and "workfare" for those with disabilities, and provide "orphanages" with a "warm environment" for abandoned children.

In *The Coming Age of Scarcity: Preventing Mass Death and Genocide in the Twenty-first Century* (1998) edited by M. N. Dobkowski and I. Wallimann, several chapters anticipate and provide plans for a sustainable society. Kurt Fintsterbusch writes: "Scarcity tends to decrease equality, integration, normative social control, democracy, and system legitimacy while increasing conflict, regulation, disturbances, repression and centralization." For a sustainable democratic society he proposes "a branch of government for planning sustainable policy, government holding corporate stocks, maximum limit to income, birth licenses, quotas for use of environmental resources, and a national land use plan." Ted Trainer suggests ways to adapt, an "alternative

economy." "Living simply . . . means being content with what is sufficient." Trainer recommends supporting the "global ecovillage
movement," envisioning a "recycling store, meeting place, barter exchange, [and] library" for the "neighborhood workshop"; the arrang-
ing of "community orchards, ponds for ducks and fish"; capital in local
banks loaned only for local projects; more communal cooperation
and less competition; control of "locally free economic enterprise"
with "social desirability"; and the "dig[ing] up [of] roads." John B.
Cobb, after recounting what is sufficient in self-sufficient agriculture,
cites Kerala in India as an example of social progress without economic growth: "In a world no longer committed to globalization of
the market, debts could and should be promptly forgiven."

SUMMARY

With differences in background and data of reference, the fertile
imaginations and perspectives of intellectual leaders reviewed in this
chapter have generated greatly different views about our future. The
diverging trend lines in Figure 3.1 represent some of these visions about
the global civilization ahead.

II

SYSTEM PRINCIPLES

In Part I we reported the state of our global system and some of the concepts, predictions, and suggestions from the many authors who have been writing about the future. In Part II we prepare for our later recommendations by giving the principles that guide all systems, including the system of economic society. Later in this book these principles of self-organization are used to understand our present condition and select successful policies.

Principles are presented in four chapters. Chapter 4 shows how successful systems of all sizes and scales use energy and materials for maximum performance. A new measure, *emergy,* helps recognize the choices that maximize real wealth. Chapter 5 explains the way systems oscillate, alternating stages of growth and descent. We try to identify the present stage of our world in the long-term pulsing rhythm. Chapter 6 defines real wealth and its relationship to money. Chapter 7 shows how systems organize and operate spatially. Chapter 8 relates population to energy and economy. Then in Part III we use the principles to select policies for a prosperous descent.

4

THE WAYS OF ENERGY
AND MATERIALS
IN ALL SYSTEMS

We start our introduction of systems principles with energy and materials. This chapter explains the way systems are controlled by energy and materials and how these vary with scales of space and time.

VIEWPOINT WINDOW

Modern science has shown that everything is connected to everything else as one universal system. However, for humans to consider, understand, or manage a small part of the universe, we can visualize an imaginary box around the parts of interest, thus defining a window of attention. For example, we draw an imaginary box around a lawn and consider the ecosystem of the grass (Figure 4.1a). To help our understanding we make a systems diagram on paper (Figure 4.1b) where the imaginary box becomes a window frame representing the system of attention. In other words, when we consider a system, we are using a window of attention to isolate a subsystem from the greater universe around it. In doing this we identify the parts and processes (the grass in Figure 4.1) and sources of inflows and outflows across the imaginary boundary (sunlight and rain in Figure 4.1).

We can adjust our mind's eye window of attention on any scale of size. For example, in later chapters we draw an imaginary line around a nation and discuss its resources in relation to its economic system, including exchanges across the boundary. Figure 4.2 shows producers, consumers, and information from consumers controlling production, and waste materials from consumers being recycled to environmental reserves available for later use. This diagram is equally appropriate for a lawn where consumers are earthworms or a nation where consumers are people.

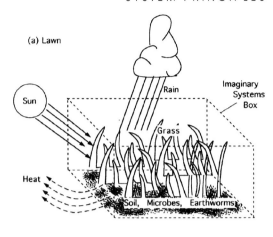

(a) Lawn

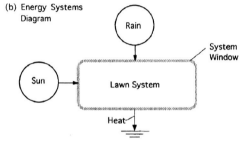

Figure 4.1. Defining a system by visualizing a window of attention. (a) Mental box placed around a lawn; (b) systems diagram with window frame that represents the box and inflow pathways for the external sources.

SYSTEM PATTERNS

When we put our window of attention on systems of many kinds and sizes we find similar patterns and relations. This makes sense because the self-organization that makes any system productive and competitive selects similar designs. Most systems contain structure, transformations, and feedback.

Storage and Structure

Instead of developing a smooth, uniform distribution of matter, self-design develops discrete units that join their functions to make a system operate. For example, our geobiosphere has organisms, clouds, rocks, lakes, cars, and banks. We can understand complex systems much easier when we represent units of a system with a few general symbols. We will explain the symbols as we go. There are symbols in Figure 4.2 for energy sources, producers, consumers, and storages. The tank symbol is used for storages. For example, materials are in a storage in Figure 4.2.

In the lawn system the producer symbol (Figure 4.2) represents the plants, and the consumer symbol represents insects, earthworms, and soil microorganisms. The producer, consumer, and box (Figure

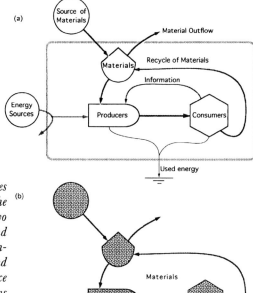

Figure 4.2. Main design features found in many systems when the window of attention includes two levels in the scale of time and space, including production, consumption, material recycle, and feedback of information service and control. (a) Whole systems diagram; (b) materials part.

4.3) are group symbols, and details may be shown within these symbols on an optional basis. For example, Figure 4.4 has typical details of the structure that is considered within a producer symbol. Notice the storage and structure (tank symbol), which receives and stores the products of plant production (organic matter, including plant structure). The energy losses from depreciation of the storage flow down the pathway that leads from the storage to the heat sink symbol. This represents the transformation of energy from a useful form into an unusable form (usually as low-temperature heat), which occurs whenever work is done.

Energy Transformations

All systems transform energy from one form to another, a process that is called "work." To do this the system must have a source of unused energy (circle symbol in Figure 4.3). Energy scientists describe energy that can do work as "available." The more available energy there is for a process, the more work can be done. After the work is done, most of the available energy has been degraded in the process—its availability used up. It is then "used energy" and cannot do any more work. Figure 4.3 shows a source of available energy, an energy transformation, an energy-containing product, and an outflow of used energy at the bottom through a symbol called a heat sink. Here the

Figure 4.3. An energy transformation process, the conversion of solar energy to chemical energy (organic matter) by plants.

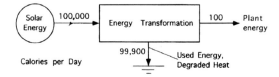

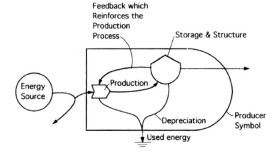

Figure 4.4. Basic design of energy transformation, storage, and feedback reinforcement found after self-organization for maximum performance.

energy transformation is represented by a box symbol, which may be used for any purpose. Degraded energy dispersed as waste heat from transformations is the lowest-quality energy on Earth. It disperses into the environment and goes out from the Earth into space. A system has to receive fresh inflows of available energy to continue operating. These universal features of energy transformations are shown in Figures 4.1b and 4.2 also.

Feedback Reinforcement

After energy is transformed, some of what is stored is fed back to reinforce the process that carries out the transformation. In the example in Figure 4.4, reinforcement happens at the symbol for interaction and production (pointed block) connecting the storage and the inflowing energy from the source. Here, after sunlight energy is transformed into green grass, the leaves spread out and capture more sunlight, thus amplifying the process with feedback action. Systems that reinforce their productive processes develop and displace those that do not. For example, by pollinating flowers, bees reinforce the processes that produce the nectar on which they feed. In the long run, farmers who maintain their soil displace those who do not feed back services to reinforce land fertility. Feedbacks of material flows and consumer services are labeled in Figure 4.2.

CIRCULATION OF MATERIALS

All systems utilize materials to make structure and store energy. Figure 4.2b shows the material-containing part of the system in Figure 4.2a. As shown, some materials exchange with the surroundings out-

Positive feedback reinforces success and survival. M.T. Brown.

side the window, while other materials recycle for reuse within the window. A system would clog up and stop working if it could neither exchange its materials nor recycle them for reuse inside. Material use requires energy to be processed, for nothing happens unless available energy is expended.

For example, chemical elements such as potassium and phosphorus are used in plant growth in a lawn and then released and recycled when the grass is consumed by grasshoppers, or dies and is decomposed by microbes and earthworms. Some of these substances flow in with water from outside (source symbol in Figure 4.2a). Although inflows fluctuate, they are balanced in the long run by equal outflows. Another example is a city that brings in food containers, consumes the food, reuses some containers, and sends out the rest as garbage.

HIERARCHY OF ENERGY ORGANIZATION

Universal Energy Hierarchy

Energy of some kind, whether chemical, wind, solar, coal, heat, or tidal energy, or energy in information, is found in everything. Any energy transformation uses up many calories of available energy of one kind just to produce a few calories of another kind. This means that all the many kinds of energy can be arranged in a series according to the calories of each kind required to make the next (Figure 4.5).

For example, many solar calories are required to make a calorie of wind, and many calories of wind energy to make a calorie of water wave energy. Another example is the food chain in an aquaculture pond. Many calories of solar energy are required to make a calorie of aquatic plants (phytoplankton algae), many calories of aquatic plants

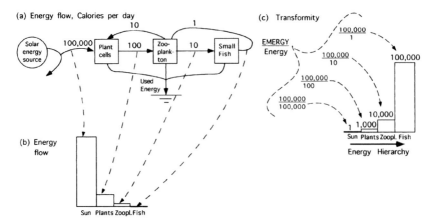

Figure 4.5. Properties of an energy transformation hierarchy. (a) Energy systems diagram of the food chain of an aquatic ecosystem; numbers are average rates of energy flow; (b) graph showing decrease in flows of available energy with successive transformation; (c) increase in transformity with each successive transformation.

required to make a calorie of animal plankton, many calories of animal plankton required to make a calorie of small fish, and so forth.

The series of energy transformations from solar energy to fishes is shown in Figure 4.5. In such diagrams that show a series of energy transformations, the ones with the most energy are on the left followed to the right by those with less-available energy after the transformations.

The word hierarchy is used when many of one kind are required to support a few of another. For example, in a military hierarchy many privates report to and support one corporal, many corporals support one sergeant, many sergeants support one lieutenant, and on up the chain of command. Where energy of each level supports less energy at the next level, it is appropriate to call the series of energy transformations an energy hierarchy. Since everything in the universe has some energy, everything can be arranged in a universal energy transformation hierarchy.

In self-organization, items that require more energy would not be produced for long if they did not have a positive effect justifying the larger resource required for their manufacture. Each transformation is accompanied by a feedback reinforcement (Figures 4.2 and 4.4). Thus a series of energy conversions has transformation pathways going to the right and feedback pathways of lesser amount but greater unit effect in reverse direction from right to left (Figure 4.5a). Self-organization generates units on different levels of energy use. Such units in hierarchical

*DILBERT © Distrib-
uted by United Feature
Syndicate. Reprinted by
permission.*

series provide a division of labor. Each level can do something to reinforce the processes of the system that the previous levels could not do. In the example in Figure 4.5, the energy of the zooplankton feeds the fish. As the fish eat the zooplankton, they control the quantity, quality, and distribution of the zooplankton population.

Another way of saying all this is:

Self-organization generates an energy transformation hierarchy.

Emergy

On Earth the most abundant source of energy is sunlight, but because it is spread out in time and space, it is low quality compared to the many other forms of energy on Earth derived from it. Many solar calories are required to make other kinds of more concentrated energy, the kinds that humans need.

It is convenient to express all other kinds of energy on Earth in terms of the sunlight energy required directly and indirectly. For this we defined a new word in 1983:

Emergy, spelled with an "m," is the available energy of one kind that has to be used up directly and indirectly to make a product or service.

We use solar emergy to compare the amounts of energy of various kinds (all expressed in solar equivalents) that have gone into the making of products and services. For energy in this book we use calories. We use kilocalories to mean 1,000 calories. The unit for solar emergy is solar emcalorie.

Transformity

The emergy of one kind necessary to generate one unit of energy of another kind is its transformity. Since we use solar emergy in this book as the common denominator for everything, we calculate solar transformities. Solar transformity is defined as the solar energy calories required directly and indirectly to make one calorie of that product or service.

> Solar transformity of a product is the solar emergy per unit
> energy (solar emergy divided by the energy). Its units are solar
> emcalories/calorie or if you are using joules, solar emjoules/joule.

For example, in Figure 4.5c 100,000 calories of sunlight energy are transformed to make 100 calories of plant energy. The solar transformity of plant production is:

$$\frac{100,000 \text{ sunlight calories}}{100 \text{ plant calories}} = 1,000 \text{ solar emcalories per plant calorie}$$

Because more energy is required to make items further to the right (Figure 4.5), they can be said to be of higher quality. Items to the right are less abundant and regarded as more valuable. Transformity increases to the right and measures energy quality.

Table 4.1 has some solar transformities ranging from sunlight, which is one by definition, to high-quality levels of information in species formation that required an estimated 1,000,000,000,000,000 solar calories (10^{15}) to be transformed. To calculate the transformity of a product we evaluate the various energies used previously directly and indirectly to make it. We express each of the contributed inputs in units of one kind of energy, solar energy. Tables like this help put items in relative importance to each other. Therefore, we can use solar transformity to indicate the position of each energy type in the energy hierarchy. Such tables make it easy to calculate the contribution of products to the economy (see Chapter 6). Details on all this are given in our book on energy accounting.[1]

Items of high transformity are worth more and can do more. Items of high transformity can be transported more easily with greater unit effect. For example, since about 4 calories of coal are required to generate a calorie of electric power, electricity has a higher transformity. Electric power is higher-quality energy, is more flexible in its many uses, and is more easily transported than coal. Electric trains are more efficient than coal-powered trains. The services of educated humans doing their jobs are even higher in the scale of transformity and usefulness of each calorie of effort.

Table 4.1—Typical Transformities

Item	Solar Emcalories per calorie*
Sunlight energy	1
Wind energy	1,500
Organic matter, wood, soil	4,400
Potential of elevated rainwater	10,000
Chemical energy of rainwater	18,000
Mechanical energy	20,000
Large river energy	40,000
Fossil fuels	50,000
Foods	100,000
Electric power	170,000
Protein foods	1,000,000
Human services	100,000,000
Information	1×10^{11}
Species formation	1×10^{15}

*calories of solar energy previously transformed directly and indirectly to produce one calorie of energy of the type listed.
Source: H. T. Odum 1996.

Transformities provide a general policy principle, which is:

Do not use high-quality products for low-quality purposes

For example, don't use electricity to heat a house if fuel in a furnace will do. Using electricity wastes the energy that was used to upgrade ordinary fuel to electricity. Don't use a highly educated worker for occupations not requiring the special training. To do so wastes the work spent in training.

Emergy Evaluation Table

For evaluating the emergy of anything it is convenient to use the values of its transformity from previous studies. As shown in the emergy evaluation table (Table 4.2), energy flow is multiplied by transformity to obtain emergy flow. The flow of emergy per time is called empower. For example, the emergy used by a nation in a year is its annual empower. The last column indicates the economic equivalent of emergy in emdollars, which are explained in Chapter 6 (see pp. 92–94).

The Maximum Power Concept

Alfred Lotka was a very original mathematician-scientist who helped start the fields of population ecology, systems ecology, biogeochemistry, and ecological thermodynamics. In 1922 he suggested

Table 4.2—Example of Emergy Evaluation Table*

Note	Item	Energy Flow Cal/yr	Transformity seCal/Cal	Emergy Flow 10^{17} seCal/yr	Emdollars[†] 10^9 em$/yr
1	Vegetables	1.02×10^{12}	2.6×10^5	2.7	0.82
2	Grains	35.6×10^{12}	8.6×10^4	30.6	9.3
3	Cattle	7.1×10^{12}	1.7×10^6	120.7	36.6

Abbreviations: seCal = solar emCalories where Cal with a capital C means kilocalories (1,000 calories); em$ = emdollars

*Production of agricultural products in Texas in 1983 (H. T. Odum, E. C. Odum, and M. Blissett 1987).
[†]Emergy flow divided by 3.3 x 10^8 solar emCalories per 1994 dollars.

the maximum power concept as a fundamental energy law rephrased here as follows:

> In the self-organizational process, systems develop those parts, processes, and relationships that capture the most energy and use it with the best efficiency possible without reducing power.

By trial and error many alternatives start to function, but only those designs that contribute more useful energy flow get reinforced and thus selected to continue. Steam engines with more efficient technology displaced the early engines that were wasteful in their use of fuels.

Designs are reinforced that maximize power output possible from the resource available. The system can then draw in more resources, produce more (Figure 4.4), and outcompete alternative patterns that reinforce less. In other words, successful systems develop structures that maximize useful resource production and consumption.

The usual system design that has production, consumption, and recycle (Figure 4.2) uses the material cycle to reinforce the capture of energy. This general design is found in biochemical reactions, weather systems, seas, geological processes, ecosystems, relationships of stars, and the human economy. The consumer unit on the right helps the producers by applying feeding services to the left. For example, trees with more leaves for catching light displace lawn grass because of the support from trunks and limbs. In this case trunks and limbs are consumers providing service to leaves in Figure 4.2.

Maximum Empower Concept

Apparently self-organizational processes take place on all scales at the same time. The rapid organization of the smaller scales is part of what supports the larger scale but is controlled by it. Organization for maximum power is observed in all scales simultaneously. The bio-

chemistry within the human body is self-organizing for maximum performance minute by minute, while the whole human is participating
in the self-organization of the economy on a much larger scale. In
each case materials and energy are being processed for maximum
production and use. Those elements or individuals whose patterns
of action do not result in maximum production tend to be replaced
eventually.

Where several levels of energy hierarchy are involved, "maximizing power" can be misleading because it might imply that priority
goes for increasing energy flow through the lower levels, which may
have more energy (for example, through plants–Figure 4.5). However, if the Lotka principle is restated as the Maximum Empower Principle, then the flows at all levels of the energy chain can be equally
suitable for reinforcement. The revised statement is:

> In the self-organizational process, systems develop those parts,
> processes, and relationships that maximize useful empower.

For example: even though the power (energy per time) through the
different levels of Figure 4.5a is very different, the empower (emergy
flow per time) is the same. The maximum empower principle accounts
for many features of system design and performance. It is a guideline
for selecting policy. For policy the principle is:

> Choose alternatives that maximize empower intake and use.

Time's Speed Regulator

Systems may operate inefficiently if overloaded. For example, an
overloaded truck can hardly move. A truck with no load goes fast, but
neither truck delivers as much per time as one with intermediate loading. Self-organization finds the intermediate loading, speed, and efficiency that maximize power. Optimum loading for maximum power
is time's speed regulator.[2]

During self-organization, selection processes develop optimum
loading on processes of all sizes within a system at the same time. In
order to refer to loading of all scales, the time's speed regulator principle is better stated:

> Systems organize loading and efficiency to maximize empower.

USEFUL INFORMATION

Useful information can be characterized as something that requires
less work to copy than to generate anew. After being selected for its
utility, duplicated, and shared, information can have great impact.
Useful information that is shared over a large territory has high

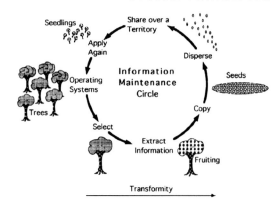

Figure 4.6. Information circle required to sustain information illustrated by the life cycle of a tree.

transformity because of the large resources required to develop and spread it. For example, the Bible and its teachings are broadly shared with a large territory, long replacement time, high transformity, and global impact.

To sustain useful information, it has to be continuously processed through an information circle as diagrammed in Figure 4.6. Copies are made (some with variations), distributed, and applied to the real systems for which the information is useful. Then after the system with the best performance is selected, the information is extracted, copied, and circulated again. In this way errors are eliminated and sometimes improvements are incorporated.

By its feedback an information circle helps to reinforce a system's performance. For example, in the life cycle of trees (Figure 4.6), the seeds duplicate and disperse the genetic information extracted from the adults. Thereafter the seed information is reapplied in the landscape to become trees. The most successful trees are selected by nature or by humans to generate seeds for the next cycle. When a material is recycled (Figure 4.2), the total quantity is unchanged, but when information is passed in its circle of support the copying process can increase the information.

SCALES OF TIME AND SPACE

Our world is made of systems of many sizes. For example, we can place the window of our attention on chemical systems of molecular size, microbial systems of microscopic size, ecosystems of small size such as plankton, ecosystems of larger size such as the forests, geological-geographic systems the size of landscapes, economic systems the size of nations, up to the universe. Within the human body are functions operating on smaller scales, but human society and its institutions operate on the larger scales, including the management of environment.

Common sense reminds us that small things turn over faster than large things. Turnover time (replacement time) is the time it takes for stored quantities to be replaced by the inflows. Tiny molecular systems turn over in microseconds, microbial systems in minutes, plankton systems in days and weeks, fishery systems in years, and society's infrastructures in decades. In recent years graphs have been made in most of the sciences, showing the way size and territory of systems correlate with those systems' replacement times. Figure 4.7 shows various systems on a graph of territory and turnover time. Small, fast systems are on the lower left and larger, slower systems are on the upper right. Scales of time and space increase upward to the right.

The windows of attention of different fields of scientific research are drawn in different places along the scale of time and space. The window for chemistry is far down to the left. The window of policy for nations is further to the right than the window for families. Individual humans starting as babies gradually move their scale of operation upward to the right as they become older and are given more territory of support and influence. Leaders draw inputs from large areas to which they feed back their influence. Because items of larger territory are usually based on more energy transformations, they have higher transformities in their flows and storages. In other words, transformities increase with scales upward to the right in Figure 4.7.

Energy-Transformity Graph

Figure 4.8 shows the basic energy hierarchy of the Earth with a graph of energy quantity plotted as a function of energy quality as measured by transformity. Since available energy has to be degraded

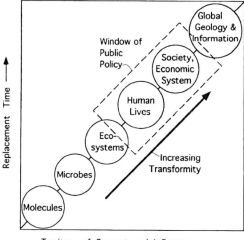

Figure 4.7. Graph showing the way turnover time, transformity, and territory of support and influence increase with different scales of time and space.

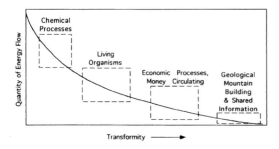

Figure 4.8. Graph of the hierarchy of energy where the distribution of quantity at each level depends on support from the level below. Quantity of energy flow is plotted as a function of transformity.

(its availability used up) in each transformation, the more transformations required to generate the higher-quality items to the right, the less energy remains. However, the energy converted is of higher transformity. For example, the available heat energy of large areas of the Earth is used and dispersed by processes that transform and concentrate a small amount of that energy in the intensive mountain building by volcanoes. Considered over the whole territory of the geologic plates, the volcanic energy is small, but its transformity is large as is its impact in organizing the system.

Because of the larger resource base and longer time of operation, the items to the right that have higher transformities can control and influence the next-smaller scales to their left. However, items may not have much effective interaction and influence with those scales more than one step away either in larger or smaller dimensions. Large fishes eat and control smaller fishes but affect the small plankton only indirectly. The large fish in turn are impacted and controlled by the next levels of size, the large carnivores such as seals and fishing boats.

Policy for All Scales

Policies that are conceived to improve productivity and uses on one scale also need to be beneficial to the connected systems, including those on smaller and larger scales. This is part of the maximum empower principle. For example, maximizing the profit of a business such as a logging company on one scale may not be good policy if it undermines the small-scale systems of the environment that are the basis for the business. Maximizing the profit in the harvesting of fish is poor policy if it undermines the food chain and recruitment stocks– the smaller-scale system on which the fishery depends.

Similarly, the policy for the current window of attention is ineffective if it does not sustain the larger-scale systems as well. Maximizing the current profit in harvesting fish is not good policy if it does not control the density of fishing boats over the seas, or omits the research and education necessary for continuing the system in the next generation (i.e., the information in the large-scale system).

SUMMARY

People can place the window of their attention on any part of our interconnected universe, using symbols to define a simplified view of one or more scales of time and space. Such subsystems have common designs that come from their self-organizing toward maximum performance. The resulting designs have consumer reinforcement of production, material recycling, and a structure of hierarchical energy transformations. Solar transformity indicates the position and importance of anything in this hierarchy. Useful information has high transformity because it requires much maintenance emergy to operate duplication and testing cycles. Emergy and its rate of flow (empower) measure the contributions of environment and economy. Policies and plans succeed by fitting with these natural principles of structure and function. Designs that maximize empower prevail.

5

PULSING
AND THE GROWTH CYCLE

Studies of many kinds of systems in many fields of knowledge show that *pulsing* is usual.[1] It appears to be a general principle that pulsing systems prevail in the long run, perhaps because they generate more productivity, empower, and performance than steady states or those that boom and bust. Pulsing of large-scale components occurs slowly, while the smaller parts are oscillating more rapidly. This chapter explains the stages of pulsing cycles, including our society, with the help of a systems model.

PULSING PARADIGM

The traditional view in ecology and the most popular concept for the future economy is growth followed by leveling off into a steady state where inflows balance losses (Figure 5.1). The process by which ecological systems such as forests grow while adding species, complexity, and diversity is called *succession*. This model is used by many in our society who seek "sustainability" for our civilization. However, the principles of pulsing apparently apply to not only environmental and economic systems but also the global surge of our civilization. An alternation between times of gradual production and storing of reserves is followed by a short period of intensive consumption and recycling. In other words, systems that produce, consume, and recycle usually pulse with patterns like that in Figure 5.2. That graph was generated with a computer simulating the relationships modeled in Figure 5.3.[2]

For example, farmers obtain maximum production by growing their pastures for a time, then grazing a herd of cattle on them for a brief, intensive period of consumption before moving the cattle elsewhere so that the cycle can repeat. Schools of plankton-eating fish move through the sea with pulsed consumption of the "blooms" of

Classical View of Sustainable Succession

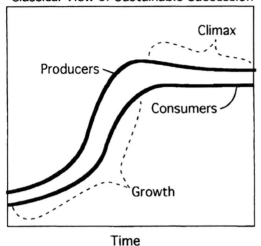

Figure 5.1. Classical view of climax and sustainability in which growth is followed by steady state.

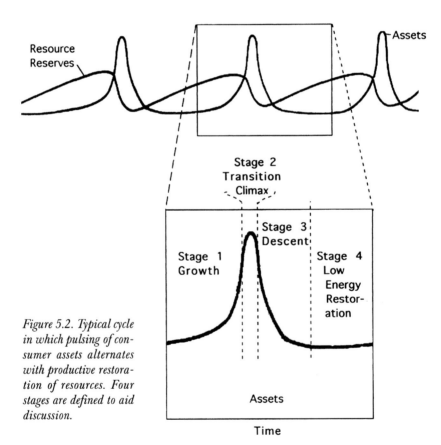

Figure 5.2. Typical cycle in which pulsing of consumer assets alternates with productive restoration of resources. Four stages are defined to aid discussion.

Figure 5.3. Systems model of pulsing production and consumption, which alternates between a time that restores resource reserves and one that uses the reserves for a surge of consumer development (pattern in Figure 5.2).

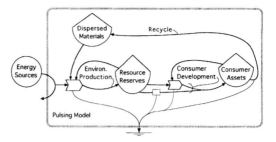

Pulsing Model

Energy flowing in from the source on the left combines with dispersed materials and resource reserves to generate environmental products that are stored as more resource reserves. Some of that storage flows out in dispersal processes, and some is transformed into consumer assets. A feedback from consumer assets helps develop more consumer assets from resource reserves. Materials released from the depreciation of consumer assets recycle and disperse.

Explanation of symbols: circle = outside source; tank symbol = storage; pointed block symbol = interaction of inputs in a production process; pathway with small block = flow in proportion to storage.

microscopic organisms that develop in a surge of growth between their visits. Pulsing is the proven norm in the biochemistry of the body. Human life is adapted to a pulsing alternation of sleep and intensive activity. Many chemical industries are more productive with batch processing than with steady state operations. For the same light energy, plant photosynthesis is most efficient when the light is pulsed. Trees grow, fall, and regrow from seedlings. In Arctic ecosystems, plant production alternates with surges of consumption by animals. The atmosphere pulses with storms. The Earth pulses with earthquakes and volcanic eruptions. Stars and galaxies pulse.

Pulsing prevails because operations that pulse transform more energy than those at steady state. Apparently, an alternation of production and consumption provides a better long-run coupling of energy intake for maximum empower than a steady state can provide. Pulsing patterns result from self-organization. Mathematical chaos is a special kind of oscillation in which quantities jump up and down and the pattern of oscillation itself oscillates. Chaos may be prevalent in many systems because the surges convert more resources into useful work.

A pulsing system of some form appears to be the stable one in the long run, repeating its periods of storage and use. There is an optimum frequency for maximum performance.[3]

Pulsing Systems of Humanity and Nature

The principle for benefiting both the economy and environment is for human systems to fit the timing of environmental oscillations.

The annual cycle of plant growth is often synchronized with the seasonal cycle of organisms that can use them. Birds adapt their times of breeding to the pulses of available food. Successful economies are those that can adjust their periods of growth to pulses in their resource basis.

There are abundant examples of the pulsing pattern of human cultures in their use of environment. Shifting agriculture was one of the common ways of alternating pulses of intensive use with an interim period of environmental regrowth and soil restoration. Anthropological studies of tribal life in New Guinea cited the alternation of periods of growing pigs, which ended with frenzied consumption activities, feasts, and intertribal war. Nomadic people timed their migrations with environmental pulses of which they were part.

Judging from history, human cultures contain the ability to switch from a regime that uses up stored resources to increase population, technological innovation, and civilization, to a quiescent regime in which the environmental reserves of forests and soils regrow. This is one of the theories for the 500-year cycle of the rise and decline of Mexico's Mayan civilizations. A low-energy period when there was regrowth of forests and thin tropical soils may have alternated with a short period of development of civilization based on intensive, unsustainable use of those resources.[4] Human societies may carry management styles for various stages of oscillation in their cultural inheritance. Figure 5.2 shows how times of consumption-based growth of consumer assets alternate with times of more gradual net restoration of resource reserves.

In this century, the rise of our capitalistic-driven civilization is one great frenzied pulse, transforming the world's resources into the assets of society. Figure 5.2 suggests how our extreme growth pulse may be one cycle among many repeating oscillations in the scale of longer view.

Pulsing Mini-model for All Scales

The pulsing curves in Figure 5.2 were generated by computer simulation of a production-consumption-recycle mini-model in Figure 5.3. The essentials for a pulsing system are an energy source, production and storing of a product (resource reserves), and a consumer-assets system that has thresholds for surged growth and decline. The surge accelerates use of the reserves, builds assets, uses the assets to consume more, and returns waste materials to production again. In principle, this model and its pulsing graphs apply to systems of many sizes from the scale of chemistry to that of the stars.

The model in Figure 5.3 is for one scale of space and time. For systems on different scales, the same model is calibrated with different turnover times appropriate to those scales. The resulting oscillations have the same shape but different time scales (some in seconds, others in years or millions of years). Note the feedback reinforcement loops (autocatalytic) already discussed in Chapter 4 (Figures 4.2 and 4.4).

As applied to our geobiosphere, slow production represents environmental processes accumulating stored reserves of minerals, fuels, and biomass, next followed by a surge of assets-stimulated consumption, which represents our urban civilization. Unfortunately, computer simulations are not yet very helpful in predicting the future because we don't know what turnover time represents our civilization. The longer the turnover time of the urban consumer assets, the longer it takes to descend as resources become scarce.

Infrastructures of our cities (highways and utilities, for example) have a turnover time of fifty to 100 years but there is much more to human civilization than its architecture and transportation. Shared culture and information best represent a civilization, and these have the largest territories, turnover times, and transformities. We can infer from the build-up and decline of the Roman civilization that a turnover time of several hundred years was appropriate then.

Simultaneous Pulsing on Many Scales

Our global system comprises systems of many scales. Each scale has pulsing with a different time period (Figure 5.4). Each of these systems is connected to larger- and smaller-scale systems so that the pulses on one scale are affected by the next scales larger and smaller. Often the small-scale oscillations, when viewed from a large-scale window of time and space, are regarded as "noise," random and irrelevant. For example, oscillations in the work of individuals in a labor force are averaged out when the work of a whole industry is considered.

On the other hand, the pulsing of a larger-scale system of high transformity can be catastrophic to the smaller scale, often requiring the small scale to reorganize and start over. For example, large-scale earth phenomena such as floods, hurricanes, and earthquakes cause the smaller-scale economies and ecosystems in their areas to repair extensive damage and reorganize. But well-adapted systems carry the means to return to function quickly as a "Phoenix rising from the ashes."

The larger-scale systems, being further up in the energy transformation hierarchy, receive less energy per area. However, by processing and storing over longer periods, they can deliver an even sharper pulse with wider influence than the smaller scale systems. For example, areas of earth movement accumulate energy over long periods before using that energy in sharp earthquakes. Energy from large areas is converged for sharp pulsing use by armies, meteorological storms, and public opinion preparing for political change. They deliver energy of high quality (high transformity).

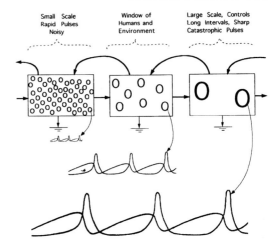

Figure 5.4. Sketch showing the pulsing for three scales of time and space. Larger scales have sharper pulses and longer interpulse periods.

Randomness or Pulsing of Smaller Scales

Many, perhaps a majority, of scientists believe there is an inherent randomness in everything from molecules to the stars. The science of statistics developed mathematical curves that describe variation. Indexes from these curves are used in experimental studies to determine whether a result is more than could have occurred considering the normal variation for that situation. Since observed variation is often interpreted as inherent randomness, many assume, philosophically, that science is limited by an inherently *indeterminate* universe. Particularly, when the scale of environment and human economy is considered, so much confusing detail is seen, that it is easy to think of it as indeterminate and unknowable.

As reviewed by D. J. Bennett in 1998, many in the history of mathematics and science believed that patterns of variation were subject to material law, with mechanistic causes. They did *not* believe in randomness. We already discussed in Chapter 4 the way everything we know about in the universe can be arranged in an energy hierarchy of many levels, each level differing from the next in territory, replacement time, and transformity. If units of all these smaller dimensions are pulsing, then it is almost obvious that the variation at any one level is due to the pulsing of the many smaller and faster components within. The distribution of variation comes from the hierarchical nature of energy. The patterns are dynamic (mechanistically caused), not indeterminately random.

POLICIES FOR EACH STAGE OF GROWTH

Like many authors before us, we can consider policy for four stages of the growth cycle shown in Figure 5.2: (1) *Growth,* (2) *Climax and transi-*

FOR BETTER OR FOR WORSE by Lynn Johnston. Distributed by United Feature Syndicate. Reprinted by permission.

tion, (3) *Descent,* and (4) *Low-energy restoration.* What is appropriate during one stage may be poor policy in another stage. For example, for a system in a stage of descent, it will not be good policy to foster growth that is no longer possible. Following is a summary of the stages of the cycle:

(1) Growth Stage

During periods of rapid growth the units of a system such as an ecosystem or a regional economy compete for resources. A few prevail by outgrowing their competitors. At this stage allowing competition to run freely insures the fastest possible growth of the whole system. Competition at this stage maximizes growth performance but inhibits diversity and drains its resources.

For example, in early stages of succession ecosystems develop rapidly with competition among weed species that increases biomass. Colonization on an economic frontier has rapid development of businesses in intense competition, with only a few predominating. Other aspects being equal, those consumers that start first tend to outgrow and overrun others. In boom towns, the people who get a head start earn money to reinvest, buy out, and dominate further development. Those that borrow get a faster start and overgrow competitors. Capitalism prevails.

(2) Climax and Transition

The climax state is a transition to the stage of coming down. Originally, ecology's word *climax* was used for the steady state (Figure 5.1), but here we use the word for the summit of development. Growth of a system continues until the quantity of structure and complexity uses up available resources as quickly as they become available. Then growth has to stop. The system reaches its maximum development, or climax. In this stage maximum performance switches from being the most rapid exploiter of resources to being efficient. This stage is sometimes said to be "mature." The main products are not growth or reproduction but maintenance and interunit services. The system becomes hierarchically organized to feed reinforcements back to the production processes.

For example, mature ecosystems tend to develop specialization in the division of labor, build more durable vegetation, do more maintenance and replacement, and recycle more materials. *Diversity and complexity increase.* Species with symbiotic, cooperative relationships develop. There is more organization. Organisms divide their tasks rather than compete. For example, specialized insects use different plants for their food, thus avoiding destructive competition. Mature coral reefs and rain forests are well-known for their high diversity. Some growth species and seeds are retained in a mature system where they are available later for the regrowth stage. Some competition among individuals of the same species continues, but competition among different species is less. In society, occupations are analogous to species. One of the open questions is how much competition is healthy for a mature stage of a human society?

A mature urban economy is similar to a mature ecosystem with many kinds of occupations, specialties, and organizations. The availability of specialists makes the life of city inhabitants more efficient. Regulation helps eliminate destructive competition. The kinds of occupations change less from year to year.

At climax there are smaller fast oscillations among smaller units, as shown in Figure 5.4. Combining the pulses of several scales generates a graph of variation like that typically observed and often called *noise.* On each scale there are pulses, even though there is a balance between new construction and removal, when averaged over a longer time. As we move into the mature stage of our civilization even now, we observe the shorter-term pulses of urban renewal, economic cycles, and political oscillation.

Mature systems contain large concentrations of information. The more complex a system, the more parts and interactions it has and the more information it must store in order to operate. In mature stages, ecosystems have high levels of information in genetic memories, neural processes of the animals, and patterns of interspecies organization. In mature stages of primitive human societies information was accumulated in the genes of their population and in the wisdom of their elders' experience. Our present human economy at maturity has great

BEETLE BAILEY by Mort Walker. Reprinted with permission of King Features Syndicate.

stores of information in libraries, universities, educated people, complex technology, and computer networks.

(3) Descent

Although history and ecosystems give us clues, we really don't know what the policies should be for the period of turndown from our complex, intensive, locally affluent, urban civilization. Assets decrease, either because the pulse of growth has used up the storage of available resources or because there is a surge of destruction by the pulse of the larger scale. With less energy, systems can only be sustained if diminished. By one means or another, the developed system has to adapt to coming down. From the chronicles of history (see Chapter 3) coming down can be gradual or catastrophic.

An unresolved question is when is it good policy to downsize gradually and when should it be catastrophic. In some systems, the mature stage is abruptly terminated by catastrophic removal due to pulses on a larger scale. The mechanisms include fire, storms, earthquakes, volcanic eruptions, hurricanes, and epidemics of insects and disease. An observer seeing a forest consumed by fire, storm, or insect epidemic thinks of it as a catastrophe. Yet in the larger scale of things this may be regular, normal, and efficient renewal.

In some other ecosystems such as a temperate forest approaching a winter season, decline is more orderly. Deciduous trees lose their leaves, animals migrate or hibernate, and consumption is reduced, but the prosperity of nature is not affected in the long run. The forests are ready to produce more leaves when the seasons change and the available energy increases again. Storages and information are set aside to facilitate a fast regrowth when resources permit—when spring comes.

Human society can adapt to diminishing resources in ways analogous to ecosystems. To sustain standards of living, populations will have to decrease. People will either adapt because of foresight or will be forced to because of the declining resources. Some people can migrate to areas where the cycle is in a different phase. For example, in rural areas of developed countries where farming was abandoned in earlier centuries, enough soil and forest restoration has occurred to support rural life again. It will be efficient to use up assets that no longer can be supported. For example, after some population decrease, materials from excess housing can be used to maintain the rest.

New specialties concerned with decrease may develop such as "downsizing business management," and these can operate for a small profit. However, choices for coming down that involve environment should be evaluated for public benefit, just as with alternatives for expansion, selecting those that sustain the most real wealth. Already we have an increase in the number of small businesses that restore and resell abandoned household appliances. More old houses are being restored. Garage sales are everywhere.

After repeated cycles of growth and decline, ecosystems develop means for carrying forward information, in seeds, eggs, and spores, for the next growth cycle. Something similar is needed in downsizing of civilization. In preparation for low-energy periods, information and diversity need to be stored in ways that will minimize loss. To maximize performance over the long term, needs of the larger scale of time should take priority. However, activities on a small scale can continue surges of growth and replacement within the overall net decrease at the larger scale.

(4) Low-Energy Restoration

After the descent, and before there can be another worldwide pulse of consumer growth, there has to be a period in which resource reserves of fuels and minerals are rebuilt. Processes of *environmental production must exceed consumption.* As consumption by the human economy declines, environmental processes of production can begin to make net storages of resources again. For this to happen, populations must decrease and adopt low-growth attitudes. It will benefit society to develop mechanisms that inhibit large-scale growth initiatives. Consumption might be limited if the information about the ways of growth is mothballed. Growth initiatives can be limited by reprogramming the cultural beliefs about growth. For example, the population crash after the potato famine in Ireland was followed by a century with religious customs that helped maintain a smaller, stable population. Policies about population and development appropriate for low-energy restoration may be like those formerly found in low-energy cultures like the Yanomamo Indians of Venezuela.

During the time when the main global economy is passing through a long down phase toward fewer people, small areas of the Earth may experience short pulses of growth and descent based on locally restored soils and forests. Where environmental production can accumulate such resources as wood, soils, and peat, local growth can occur until exploitation uses up the accumulations. Then conditions are ready for a new cycle of resource accumulation, growth pulse, and consumption that in turn recycles the materials for another time of net production and restoration.

Agrarian cultures can operate at lower levels with shifting land rotations of twenty to 500 years. This way of life could combine some modern knowledge—for example, genetics—with crop varieties saved from earlier agriculture that require fewer fuels, fertilizers, and chemicals. Geological processes continuously bring small quantities of minerals and fuels to the surface locally in reach of human use. Small pulses of development can occur using these resources more frequently around mountains and river deltas where earth cycles are rapid. But for the lower-energy future of the Earth as a whole, we can expect dispersed smaller-population communities living primarily on renewable energies like wood, crops, fish, and hydroelectric power with some use of the slowly renewable fuels and minerals.

Lights of Pulsing from Satellite

The hierarchical centers of consumption in cities have had sharp pulses of growth and decline over history. If a person could have watched from orbit over centuries, the pulses of light rising and falling first in one place and then in another would be like a Christmas tree with flashing lights. Looking out to space over eons, there are the even longer periods of pulsing of even greater centers of energy use, the stars.

ONE SURGE VIEW OF CIVILIZATION

Perceptive energy realists (M. K. Hubbert 1949; E. F. Cook 1976; C.A.S. Hall, C. J. Cleveland, and R. Kaufman 1986) have long described our civilization as a flash in the pan when viewed in the longer scale of human existence. According to the model accepted by many resource scientists, the high development of the present civilization depends on fuels and minerals previously stored by Earth processes over a long period of time. Although "nonrenewable resources" are being slowly renewed by Earth cycles, a very long time is required to bring enough fuels to the surface for another pulse like the nineteenth and twentieth centuries. The present consumption by society of the storages near the Earth's surface is much faster than the rate of restoration by the cycles of the Earth. Hence they are called "nonrenewable" resources so far as our current civilization is concerned.

Later developments have to depend on the renewable energies. These include the tidal energies, the heat of the Earth, and especially the solar energy that drives the winds, waves, and biological productivity of nature, agriculture, forestry, and fisheries. Small, short-term surges of development based on Earth products can develop in local areas.

Renew-Nonrenew Model

For computer simulation a simple model based on renewable and nonrenewable resources is provided in Figure 5.5. It does not have pathways to refill the stored reserve and thus does have the repeating oscillations in Figure 5.3. Figure 5.5b is a typical simulation of growth and descent of the simpler model. First assets of civilization (on the right in Figure 5.5a) grow slowly with an agrarian production economy that uses renewable sunlight, rains, and sustainable soil. Eventually the assets surge into unsustainable growth when technology is found for rapid use of the nonrenewable resources (fuels and minerals). After the one surge of growth, assets return to a lower level based on the

Figure 5.5. Renew-Nonrenew *model for the burst of civilization as it consumes a nonrenewable fuel reserve. (a) Systems diagram; (b) typical result of computer simulation.*

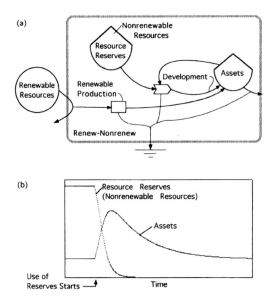

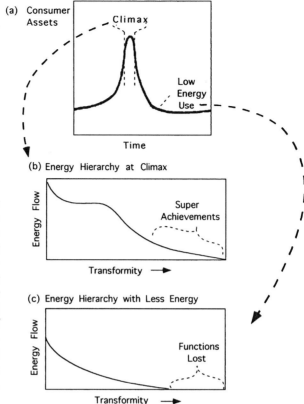

(a) Consumer Assets

Climax

Low Energy Use —

Time

(b) Energy Hierarchy at Climax

Energy Flow

Super Achievements

Transformity →

(c) Energy Hierarchy with Less Energy

Energy Flow

Functions Lost

Transformity →

Figure 5.6. Graphs of energy distribution with transformity that compare the energy use hierarchy at climax stage with that at a lower energy stage. (a) Stages of growth and decline; (b) achievements at climax; (c) achievements less in times of lower energy.

renewable energies again. The renew-nonrenew model represents one period of time within the longer regime of repeating pulses produced by the more complex model (Figure 5.3). This model shows simply the "limits to growth" concepts of many authors cited in Chapter 3.

Hierarchy with Descent

At times of high available energy (the climax stage in Figure 5.6a) more work can be done, and higher levels of hierarchy can be developed. Large volumes of lower-quality energy use can support more high-transformity developments. Later the quantity of achievement at high transformity has to diminish when the systems use less energy.

In Chapter 4 we introduced an energy-transformity graph (Figure 4.8) to show energy hierarchy and relate quantity to quality. Compare this graph for the time of climax (Figure 5.6b) with the one shown for times of lower energy (Figure 5.6c). There is a loss of the top functions (high transformity on the right), because the energies at the supporting levels below are less. Examples of high-transformity items are the space program, powerful military weapons, bureaucracies of political leaders, many tiers of celebrities, worldwide social movements, and the extreme developments and storms of information.

SUMMARY

Most systems pulse, alternating between a time of product accumulation and a time of rapid autocatalytic consumption that transforms the reserve into temporary, high-quality assets. We described the cycle of assets in four stages: Growth, Climax-transition, Descent, and Low-energy restoration. Our civilization seems to be in the climax-transition stage with turndown ahead. Later in the cycle the energy flow is less, the hierarchy shorter, and transformities less. Pulsing oscillations occur simultaneously on all scales, mutually coupled. The small-scale pulses appear as noise, while the large-scale pulses are catastrophes. Computer simulating a generic model generated typical pulsing patterns. A simple one-surge model of our civilization (*Renew-nonrenew*) dramatizes simply the essence of the transition and descent ahead.

6

REAL WEALTH AND THE ECONOMY

Food, shelter, clothing, fuels, minerals, forests, fisheries, land, build-ings, art, music, and information are *real wealth*. Money by itself is not. Money is circulated among people who use it to buy real wealth. Whereas money measures what people are willing to pay for products and services, *emergy* (see Chapter 3) measures real wealth, including both work contributed by environmental systems and that contributed by humans. Money and markets function well enough on the scale of size and time of human lives, but are inappropriate on the smaller scales of environment and the larger scales of geology and global in-formation. This chapter explains principles that relate resources, real wealth, and the economy, and are the basis for policies in chapters that follow.

MONEY AND THE EARTH SYSTEM

Figure 6.1 shows the chain of energy transformation on the planet Earth, starting with the energy inputs on the left, which lead up to the people and information on the right. Money circulates among people in the middle scale of humans and their business enterprises. Figure 6.1 shows how money (drawn with dashed lines) is a countercurrent flowing in the opposite direction of, and in exchange for, the goods and services it buys. Although the real wealth of the world economy is not money, the smooth circulation of money helps maximize global performance by accelerating processes in the medium scale of human business and commerce.

Although people pay money to each other to acquire the real wealth products from the Earth's environment, the circulation of money is not involved directly in the basic environmental systems of produc-tion of real wealth (Figure 6.1, on left). Although there have been

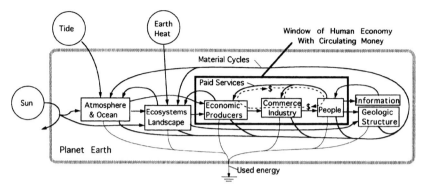

Figure 6.1. Systems diagram showing the basis for economic buying power in the energy hierarchy of the Earth. Dashed lines indicate the circulation of money.

attempts in recent years to increase control of information for profit through patents and copyrights, much of the information of the planet (Figure 6.1, on right), including the genetics of life and the learned information of society, also operates and circulates apart from money. For example, the biodiversity of the Earth, public opinion, births, and family education are part of the production and use of real wealth but not controlled directly by markets. The great Earth cycles of land and mountain building are also larger than can be controlled with markets.

GLOBAL WEALTH

Let's summarize the relationships of resource use and circulating money with a window of attention on the whole Earth (Figure 6.2).

The Emergy-Money Ratio

The circulation of money for production is often measured by the gross economic product in dollars per year. The total use of resources to produce wealth is measured by the emergy use in solar emcalories

HAGAR THE HORRIBLE by Dik Browne. Reprinted with permission of King Features Syndicate.

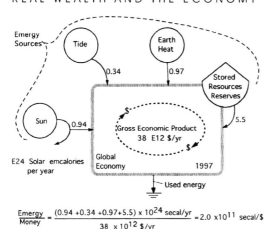

Figure 6.2. Diagram with the window of attention on the global circulation of money and its source of real wealth from current energy sources and environmental reserves. Emergy-money ratio was defined to indicate the real wealth (emergy) buying power of money, with worldwide values from M. T. Brown and Sergio Ulgiati (1999).

$$\frac{Emergy}{Money} = \frac{(0.94 + 0.34 + 0.97 + 5.5) \times 10^{24}\ secal/yr}{38 \times 10^{12}\ \$/yr} = 2.0 \times 10^{11}\ secal/\$$$

per year. When the global emergy use is divided by the money circulation, the quotient–the emergy per unit money–measures the average buying power of money worldwide. The emergy-money ratios for various countries are given in Table 6.1. These were calculated by dividing the yearly emergy use in a country by its Gross National Product (GNP) in international dollars. (Circulation of money in each country was converted to U.S. dollars according to the exchange rate at the time.)

Notice in Figure 6.2 that about one-third of the emergy use currently comes from renewable resources and two-thirds from the nonrenewable resource reserves such as fossil fuels. If the nonrenewable resources were no longer available and the circulating money was the same, the ability of money to buy real wealth would be reduced by half. Since money is paid only to people, the emergy-money ratio measures the average real wealth exchanged for human service.

Emergy Standard of Living

If emergy use measures the real wealth, a good index of standard of living is the annual emergy use per person (found by dividing a nation's annual emergy budget by its population). This index includes the real wealth that rural people derive directly from the environment without spending money, something the dollar-based cost of living index can't do. Budgets of emergy use and population density vary widely from one area to another. Table 6.2 shows the emergy per person for different countries.

Emdollar (Abbreviated Em$)

As illustrated with Figure 6.2, the total budget of circulating money measured by the gross economic product buys real wealth as measured in emergy units for the year. Emdollar evaluation of something

Table 6.1—National Activity and Emergy-Money Ratio*

Nation	Emergy used $\times 10^{22}$ secal/yr	Gross National Product $\times 10^9$ $/yr	Emergy/Money $\times 10^{12}$ secal/$
Liberia	1.1	1.34	8.2
China	171.8	376.0	4.6
Dominica	0.017	0.075	2.3
Ecuador	2.2	11.1	2.1
Brazil	42.6	214.0	2.0
India	16.1	106.0	1.52
Australia	21.1	139.0	1.52
Poland	7.9	55.0	1.44
Former Soviet Union	103.0	1,300.0	0.79
U.S.A.	199.0	2,600.0	0.76
New Zealand	1.88	26.0	0.72
West Germany	41.8	15.0	0.58
Netherlands	8.8	166.0	0.53
World	554.0	11,600.0	0.49
Spain	5.0	139.0	0.36
Switzerland	1.75	102.0	0.17
Japan	36.6	3,060.0	0.12

secal = solar emcalories

*Values are from different years in the 1980s (H. T. Odum 1996, Table 10.7).

is the dollar share of the dollar budget (gross economic product) estimated from its emergy share of the whole emergy budget. For example, if the agriculture of a nation contributes 20 percent of the total annual emergy budget, then 20 percent of the gross economic product is from agriculture. The annual emdollars from agriculture is 20 percent of the gross economic product. This is larger than just the dollar value, because the dollars paid for products are only for the human services involved.

The *emdollar,* abbreviated *Em$,* is the emergy contribution that goes to support one dollar of gross economic product—after evaluating the emergy flow of something, divide by the emergy-money ratio (solar emcalories/$) for that year and country to obtain the contribution in emdollars of that year. We often use the last column in tables of emergy calculations for emdollars (Table 4.2).

ENVIRONMENT-ECONOMIC INTERFACE

Next we zoom our window of attention down to the relationship of environmental resources and economic use. Economic use begins when humans begin to harvest and process environmental products and

Table 6.2—Annual Solar Emergy Use per Person*

Nation	Emergy used x 10^{19} secal/yr	Population x 10^6	Emergy Use per Person x 10^{12} secal/person/yr
Australia	21.1	15.0	14.1
U.S.A.	198.0	227.0	8.7
West Germany	41.8	62.0	6.7
Netherlands	8.8	14.0	6.2
New Zealand	1.88	3.1	6.1
Liberia	1.1	1.3	8.5
Former Soviet Union	103.0	260.0	4.0
Brazil	42.5	121.0	3.5
Japan	6.5	121.0	3.0
Switzerland	1.75	6.4	2.7
Ecuador	2.3	9.6	2.4
Poland	7.9	34.5	2.3
Dominica	0.017	0.08	2.1
Taiwan	3.2	17.8	1.8
China	171.2	1,100.0	1.55
World	554.0	5,000.0	1.11
Spain	5.0	134.0	0.37
India	16.1	630.0	0.26

*Values are from different years in the 1980s (Odum 1996, Table 10.8).

services. Examples include agriculture, forestry, fisheries, and the visiting of parks by tourists. The general plan for economic use in Figure 6.3 shows the products and services going from the environment to the economic users, where money circulates. The economic user harvests, processes, and transports the environmental product to the rest of the economy, on the right of the diagram, receiving money in exchange. The money is paid back to people in the economy for necessary inputs. No money is paid to the environment for its extensive work.

The money paid to the processors for the environmental products is not an indication of the amount of real wealth being contributed by resources to the economy. In fact, costs and prices are often inverse to the environmental contribution. When fish, forest wood, crops from rich soils, and wilderness products are abundant, costs of processing are small. With products easily available, prices are low. Low market prices might imply that these products have low value (if you try to use money to measure value). But under these conditions everyone has abundant real wealth, money goes a long way, standards of living are high, and human populations will grow.

*Figure 6.3. An interface
between the environment
and the economy showing
local free contributions
(from the left) and inputs
of resources and human
services purchased from the
main economy (to the
right). The dashed line is
the flow of money.*

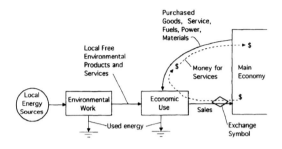

Later, when resources become overused and scarce, costs and prices are high, and high market prices imply that the products are valuable. But under these conditions the real wealth going to people is small and what they must give in return is large. The conclusion is profound:

> Prices and costs cannot be used to evaluate the contribution of
> resources to the real wealth of an economy.

In other words, money cannot be used to evaluate resources from the environment, impacts on the environment, or measures planned to help the environment, because payments go only for the human part of the work. The extensive efforts to put a market price tag on nature cannot tell us very much about what the real value of nature is. However, emdollar values do measure nature's contributions.

Inputs to Economic Use

As viewed in Figure 6.3, there are two sets of contributions to environmental products as they enter the economy: (a) the environmental contributions from the left that are free, and (b) the purchased inputs from the economy coming from the right (feedback) including human services, foods, fuels, electric power, and materials. The dashed line represents the money coming from sales of the product going up to pay humans for the input of their services.

With the local view of Figure 6.3, the environmental inputs from the left are free. However, the inputs from the economy on the right also came from the environment, but in some other place, before they were incorporated into the economy. As we explained in our worldview (Figures 6.1 and 6.2), all real wealth ultimately comes from some environmental system somewhere.

More details on the economic use of environmental products are given in Figure 6.4a. There are two kinds of purchased inputs: the commodities (examples: oil, electric power, materials) and the human services. The money goes only to humans for their part (oil company services, electric company services, food processing services, etc.). For example, the money you pay for a gallon of oil pays for workers in oil

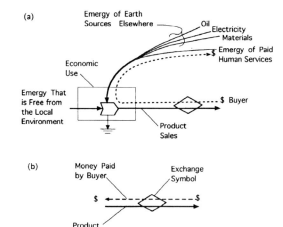

Figure 6.4. Diagram of sources of emergy in economic exchange. (a) Details of economic use of a local resource; (b) exchange in an economic purchase.

fields, those who provide financial services, and those who manufacture equipment, but no money is paid to mother nature for the most valuable part: the chemical energy in the oil.

Exchange Ratio

When products or services are bought and sold, money is exchanged as diagrammed in Figure 6.4b. Both flows can be expressed in emergy units. The emergy in the commodity for sale includes that previously contributed by all the local, free inputs and that purchased from elsewhere (Figure 6.4a).

The emergy of the money flow is that of the services it can buy. Multiply the dollar flows by the emergy-money ratio (Figure 6.2). The emergy of the flow of commodity or service is calculated by multiplying the energy in the flow by its transformity (emergy per unit energy). The ratio of the emergy received to that paid is the *emergy exchange ratio,* a useful indicator of benefit. All the resource examples in Table 6.3 have high emergy exchange ratios indicating much more real wealth received by the buyer than the seller.

$$\text{Emergy exchange ratio} \ = \ \frac{\text{Emergy of products bought}}{\text{Emergy of payment}}$$

Net Benefit to Purchasers of Environmental Products

Since money is paid only for part of the inputs (Figure 6.4a), it can represent only part of the product's value. For example, when you buy fish, you are paying for the work of the fisherman and the work of people involved all along the line in supplying equipment, fuels, etc., but you do not pay the ocean for the fish or pay the Earth for the fuels.

Table 6.3—Using Emergy to Compare Environmental Contributions and Their Market Values*

Item	Unit	1983 Price Dollars	Emergy Ratio[†] (Buyer's Benefit)
Water	Acre-foot	50.00	1.9
Potatoes	100 pounds	8.50	2.0
Fuel	Gallon	1.00	3.3
Wheat	Bushel	3.55	3.5
Cotton	Pound	0.59	3.9
Beef	100 pounds	55.00	6.5
Fertilizer	Ton	164.00	11.8
Wool	Pound	0.83	16.7

*Prices were given in *Texas Livestock, Dairy and Poultry Statistics for 1983,* Texas Dept. of Agriculture (Odum et al. 1987).
[†]Ratio of emergy in commodity to emergy of money paid.

Since buyers receive both the free contributions of nature plus the bought human services, they receive more real wealth than they pay for. In other words, there is usually a large net benefit to the buyer. In Table 6.3 typical exchanges were evaluated by putting both in emergy units and expressing as exchange ratios.

Primary Sources and Net Emergy Yield

Primary resources are those that contribute much more emergy to the economy than is required for their extraction and processing. The *net emergy yield* of a product is defined as the emergy yield minus the emergy from the economy used to process the product ($Y - F$ in Figure 6.5). Examples of products with high net emergy yields are the fossil fuels. They make a large net contribution that can be used to support other aspects of the economy.

A good measure of the net contribution is the *emergy yield ratio,* the ratio of emergy in the yield to that required from the economy for the extraction and processing (Y/F in Figure 6.5). Procuring primary resources with high net emergy yields is one of the most important policies for a nation. Prior to the energy crisis of 1973, many fuel sources with an emergy yield ratio of 40 to 1 or more were available to developed nations. These countries had the excess resources to develop rapidly, operate large military establishments, and conduct innovative research in science and technology. However, as prices of fuels have risen worldwide in the last two decades, lower emergy yield ratios (3 to 1 to 12 to 1) have prevailed, giving nations less to work with

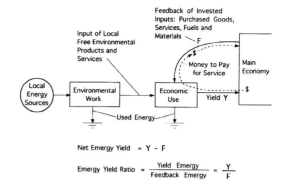

Figure 6.5. Diagram
of the interface be-
tween the environ-
ment and the economy
showing the emergy
yield ratio *used to
evaluate net contribu-
tion to the economy.*

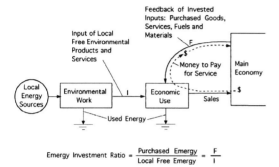

*Figure 6.6. Diagram of
the interface between the
environment and the
economy showing the
emergy investment
ratio used to evaluate
development intensity
and economic feasibility.*

and reducing potential for growth. Energy sources are discussed in
more detail in Chapter 9.

Economic Feasibility with the Investment Ratio

Most economic sectors do not have to produce net yield. They use
the net yields from primary energy sources, which they use to make
their contributions. Thus goods, services, and other inputs from the
main economy are used to process environmental resources. In Figure 6.6 note the paid inputs from the right (which we call the *invested
inputs*) and the free environmental inputs coming from the left. The
emergy investment ratio is defined in Figure 6.6 as the purchased emergy
inflow divided by the emergy of the free environmental inputs. This
ratio is useful for evaluating the intensity of the economy's contribution to an area. The average emergy investment ratio within the United
States is about 7 to 1, whereas in less-developed areas the ratio may be
1 to 1 or less. Urban areas with a high intensity of development have
greater ratios utilizing and impacting the environment more.

The more real wealth contributed free from the environment compared to what has to be purchased, the more benefit there is to the
region's economy. The more free environmental contribution, the lower

the prices can be for the product and the more successful the project is in economic competition. In other words, the lower the ratio is for a project compared to that typical of the region, the more likely it is to be economically feasible.

For example, a paper mill with an abundant source of clean water and with wetlands on its property that can accept and process the dark brown wastewater without harm, buys less and has more free service from the local environment than another mill. Therefore it has a lower emergy investment ratio and low prices for its products. A competing business that has to buy and transport water and operate additional treatment plants for its wastes buys more from the economy and gets less free from the local environment. It has a higher emergy investment ratio and the prices it has to set for its product are less competitive.

The ratio is useful for housing developments and their aesthetic-recreational support from the environment. Units in a new condominium surrounded by green areas can be sold for a high price and high profit. The green areas provide shade, healthy areas for children to play in, and a peaceful place for adults to relax. Later, if there is a large crowded group of condominiums, with all the land paved over for parking and no green areas left, the apartments are sold for less, and the services once provided by nature must be purchased.

Reinforcing Environmental Production

The free harvests from the environment shown in Figures 6.3, 6.5, and 6.6 are not sustainable in the long run because the humans exploiting them do not reinforce the environmental production systems. This is the *principle of reinforcement for maximum performance* introduced in Chapter 3. The stocks (storages) of the environmental systems are *natural capital* and are necessary for continued environmental production. Harvesting without reinforcement eliminates the natural capital. For example, harvesting a forest without replanting causes weedy species to take over the land. Economic markets accelerate degradation

Figure 6.7. Diagram of the interface between the environment and the economy showing the feedback reinforcement necessary for long-term sustainability and the money circulation (dashed lines) to pay for it.

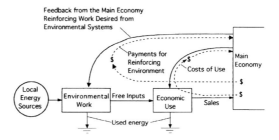

because scarcity increases prices, keeping the harvesters stripping what is left until they are out of business. Bottom fisheries off New England that collapsed because of overfishing were displaced by less-edible species such as dogfish (mini-sharks), making it more difficult for cod and haddock to be restored.

Thus to be sustainable the economic use design has to include a reinforcement from the main economy back to the environmental process (Figure 6.7). The reinforcement requires money to be paid for the human service part of the reinforcement (second dollar-flow in Figure 6.7). In agriculture this is done with fertilizer and soil conditioning. Those who wish to maintain environmental production must alternate cycles of growth and harvest. For example, crop rotation is a way to maintain agricultural soil fertility. Land use for crops alternates with periods of wild vegetation growth (a fallow stage) that restores soils. Forest cutting, if it is to be sustained, must be followed with the reinforcing work of reforestation.

However, when short-range economic competition for market prices is intense, farmers are forced to omit the long-range care of the land, causing loss of productivity. In the past, many countries provided "subsidies" from the main economy to stabilize agriculture. This was good long-range policy. An example of a government subsidy that does reinforce the land productivity is the aerial spreading of phosphate fertilizer over pastures in New Zealand. However, aid to the environmental processes should not be confused with payments to people who use the money for something else.

Policies that maximize profit to the user tend to be unsustainable because they strip the environmental base. A sustainable production system (Figure 6.7) will not make as much short-term profit, but it uses some of its emergy yield to create new jobs for long-term work on the environment.

MACROECONOMICS OF THE GLOBAL SURGE

Several authors cited in Chapter 3 considered causes of the business cycles and other oscillations in human societies. Many of the theories

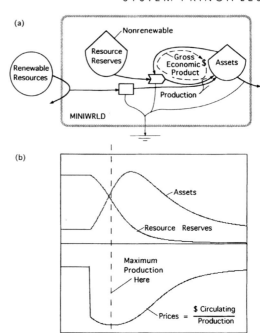

Figure 6.8. Simulation of the model Miniwrld, *a revision of the renewable and nonrenewable model of resources in Figure 5.5. (a) Systems diagram of model with constant rate of money circulating; (b) simulation results showing initial fall in prices and later rise.*

involve accumulations followed by a pulsed use. The one-surge model (Figures 5.5 and 6.8) that follows dramatizes the way macroeconomics and prices depend on the coupled pulses of real wealth.

Prices and Development

We coupled money circulation to the *Renew-nonrenew model* from Figure 5.5, and simulated the revised model *Miniwrld* with money supply held constant (Figure 6.8a).[1] The pulse of the growth and decline of assets caused changes in prices and other economic properties. At first, during early stages of growth, prices dropped as assets, including technology and information, increased (Figure 6.8b). This occurred even though the availability of stored resource reserves was declining. At first, efficiency increased because of use of the resource to build assets that autocatalytically feed back to production, just as authors like J. Simon and H. Kahn have emphasized (see pp. 50–53). Prices decreased. Maximum production occurs when resource reserves and assets are both moderate. Later, however, when the resource decline began to limit production, prices rose. Less emergy was received per dollar, and more had to be invested from the economy to produce the same product. After the resource declined further and productivity had passed its maximum, prices rose and assets turned down.

From EARTHTOONS by Stan Eales. Copyright © 1991 by Stan Eales. By permission of Warner Books Inc.

Scale for Money

Monetary systems have evolved in human cultures many times, and the survival tenacity of human monetary exchange indicates their utility. In our time many have faith in the free markets and economic incentives to manage everything. But when considered on a wide range of scales of space and time (Figure 6.1), the window of monetary circulation occupies only a middle position. As written by many authors cited in Chapter 3, important scales of our global system (Figure 6.1) are smaller and larger than what is appropriate for market control. Whereas a free market system with competitors tends to reinforce those individuals and businesses that produce more, it cannot effectively manage sectors that don't receive money and are not partners in commerce. These sectors include especially the environment (see Chapters 15 and 16) and information (see Chapter 17). Stated in another way, there is a limited range of transformities that are best managed by money and markets.

Appropriate Use of Capitalism

Many authors cited in Chapter 3 were concerned about the detrimental role of excessive global capitalism. Clearly, policy designed to make a vital economy for people is required to assist, not interfere with, the other levels of the whole global system of real wealth production. It is poor policy to maximize profit for a part of the system in the short run, if this reduces the real productivity and wealth for the whole system in the long run.

For example, on a local scale, private capital invested by large-scale logging of old-growth forests may make a momentary profit by stripping an area quickly before going out of business. However, vital properties such as tree populations and soil and water management capabilities are lost. Reducing forest capacity that could be sustainable with a smaller-scale rotation hurts the long-range local economy. The maximum empower principle (see Chapter 4) suggests that policies that do not maximize the useful work of the whole system are soon selected against. In this case, public policy that allows stripping of natural capital will be forced to change when it is found that the whole economy has been diminished. For example, constraints are now being placed on overinvestment in fishing operations.

For the world economy, many authors cited in Chapter 3 reported global capitalism with unsustainable extraction of natural capital from less-developed areas for benefit of the developed nations. In Chapter 8 we will use emergy of international exchanges and finance to understand the present global inequity in distribution of income and real wealth and suggest remedies for that inequity.

As a mechanism for feedback-reinforcing growth, free market capitalism has dominated during recent times of growth and succession, but its role may be different and less important during times of leveling and descent. People in developed countries take for granted that money is entitled to earn interest, whereas paying interest depends on growth somewhere. Buying power can't increase if there is no increase in real wealth. Adding money without growth merely inflates the money's value.

Even when developed countries are not showing much real economic growth domestically, their international capitalistic ventures can expand as long as there are real wealth resources to be exploited from the less-developed countries. Gross economic products can increase in developed centers as long as there are resources to strip abroad.

As the entire world finally becomes resource limited, the world consumption must level off and decline as in the simulation model (Figures 5.5 and 6.8). With few resources left to develop, large loans and interest cannot be repaid, and large-scale capitalism disappears. Loans and investment are then restricted to a smaller scale, replacing essential structures and developing smaller patterns for a time of lower energy. If new policies are not developed for a smooth transition, the arrival of the new era might cause market crashes and crises.

SUMMARY

This chapter relates and differentiates between money and real wealth. Circulation of money increases empower and efficiency at the intermediate scales of human society, but does not adequately measure contributions of environment or information on small and large scales. Real wealth was measured in emergy and emdollar units and related to the buying power of money with an emergy-money ratio. Emdollars indicate the part of the buying power of the gross economic product due to real wealth. World economic development accelerated by global capitalism has caused inequitable processing of real wealth and large differences in the real wealth buying power of dollars in different countries.

Simulation of an overview mini-model shows how resource prices decrease with technological development at first until there is inadequate nonrenewable energy to support them. Then turndown in the stock of real wealth raises prices again and reduces the power and scale of capitalism.

At the interface between environmental resources and economic users, money is only paid to humans for inputs of labor and services but not to nature. Emergy flow (empower) was used to value the contributions from the environment and from the economy on a common basis. Emergy yield ratio was used to determine the net benefit of using a resource. Emergy investment ratio was used to measure economic advantage of an environmental investment and its impact. These analyses provide a basis for predictions and policies for future development. Economic reinforcement of natural capital is necessary to sustain contributions of environmental systems to the economy.

7

SPATIAL ORGANIZATION

All systems develop hierarchical spatial patterns. As energy is transformed, the products flow toward concentrated centers. Materials circulate between the centers–concentrations of consumption, money, information, empower, high transformity, structure, and control–and the sparse areas. This chapter describes the concepts of spatial organization and the way the economy and environment become geographically organized to increase productivity.

SPATIAL PRINCIPLES

Production, Consumption, and Recycling

Wherever we place the window of attention we can find systems of production, consumption, and material recycle (Figures 4.2, 7.1a). Where the source of energy is entering the system evenly spread out on a broad surface, the production that captures this energy has to be broadly dispersed too. For example, fields of agriculture capture the sunlight. After energy is transformed into more valuable products, these converge spatially to be used by consumers in centers (Figure 7.1b). For example, in many agrarian landscapes before the industrial revolution, agriculture around a city supplied food for people and hay for the horses in the towns.[1] Although the energy reaching the consumers was much less than the original solar energy, it was of higher quality (for definition of *higher transformity,* see Chapter 4). The products reaching the city consumers became concentrated as they converged toward the center.

For maximum performance, the central consumer part of the system returned services to reinforce the rural system. For example, day laborers and equipment moved from the city to the farms. Consumer centers also returned the "waste" materials released during the con-

Figure 7.1. Spatial arrangement of production and consumption. (a) Systems model; (b) geographical view of energy flows; (c) geographical view of material cycle.

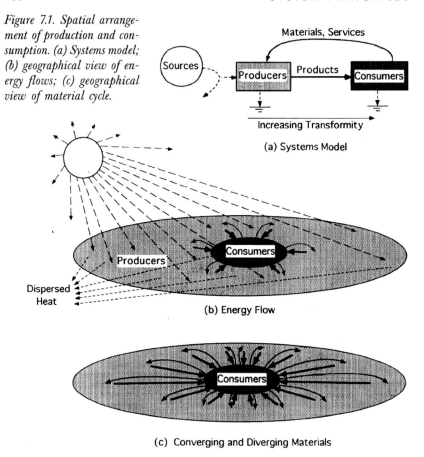

(a) Systems Model

(b) Energy Flow

(c) Converging and Diverging Materials

sumption process. For example, the agrarian city returned horse manure to the surrounding farms. Figure 7.1c shows the materials from the center flowing outward to the producers in diverging, dispersing pathways, thus closing the cycle of materials necessary for continued production. Production and consumption are symbiotic. Both are mutually stimulated by the products converging to the center and materials and services diverging back.

This converging and diverging design is observed in many kinds of systems. Cumulus clouds are the centers of converging and diverging airflows. Some habitats of animals such as oyster reefs are consumer centers that capture and concentrate the organic matter from large areas of plant production, releasing wastes that are dispersed into the surrounding area. Ants bring foods from surrounding areas into their mounds and carry wastes out. Volcanoes are centers where force and energy converge from surrounding areas, and materials are dispersed outward when there are eruptions.

A good policy for planning a landscape is to arrange for materials to converge and diverge in complete cycles. Expenditures on dispersing wastes back to the rural systems may be as important in the long run as those spent to bring products into the centers of an economy. For example, the organic wastewaters from paper manufacture need to go back to pine forests after forest products are sent to the mill. Urban transportation is organized to converge people to the center of cities and recycle them back to their dispersed homes again.

CALVIN AND HOBBES © *Watterson. By permission of UNIVERSAL PRESS SYNDICATE. All rights reserved.*

Many Levels of Hierarchy

With the production-consumption model, the window of our attention was on two levels of hierarchy (Figures 4.2, 5.3, and 7.1). But the real world has many scales of size and time (Figure 4.7). Each level receives products converging from smaller units spread out over larger areas. Each level in turn creates products for the more concentrated center at the next higher level.

Whether you use the words "producer" or "consumer" depends on your window of attention. For example, within an ecological window of interest, fish are consumers receiving energy from smaller, more dispersed producing units—plankton in the sea. The window of attention for a family living off fish sees the fish as the producer and the family as the consumer. On yet a larger scale with a window of attention on economic use, fishermen are the producers and the people in the towns are the consumers. The energy chain in Figure 4.5 shows the way the energy decreases through successive levels but the transformities of the products increase. Here sunlight supports phytoplankton, which support zooplankton, which support small fish.

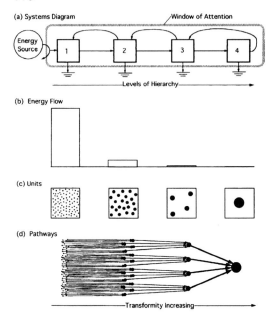

Figure 7.2. Properties of a system with four levels of hierarchy. (a) Systems model; (b) energy flow; (c) numbers, sizes, and territories of units; (d) energy flows passing up the hierarchy. Feedback flows can be represented similarly but with arrows reversed (not shown here).

Four levels of energy hierarchy and size are represented in Figure 7.2 to show the spatial pattern that emerges. From left to right in Figure 7.2c, units get larger and fewer with larger territories of influence, longer turnover times, and higher transformities. Examples are the food chains of aquatic ecosystems, the ecological organization of land ecosystems, and the spatial organization of farms, villages, and cities in an earlier agrarian economy. As with a two-level window, the transformity increases with each step from geographically dispersed small units on the left to more centralized units on the right. The total energy flow declines, but the converging pattern brings highly concentrated flows of high emergy into the centers. Figure 7.2d shows the convergence of the pathways passing up the scale of hierarchy. The reinforcing, return feedbacks omitted in Figure 7.2d follow similar patterns except with arrows in reverse direction. Spatial divergence of feedbacks is shown in Figure 7.1.

Gaps

Wherever an area is in a low-energy, restoration period of its growth cycle, it appears as a gap in the better-developed areas around it. For example, wherever trees have recently fallen in a forest, there is a gap in the cover of leaf canopy. Studies in geography, landscape ecology, oceanography, and other fields have observed the geographical property of many small gaps and few large ones. Because of many rapid oscillations in small units and fewer slow oscillations in larger units,

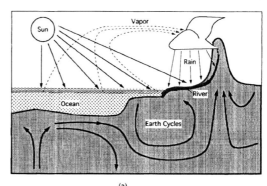

(a)

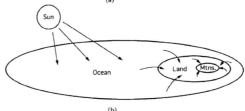

(b)

Figure 7.3. Biogeosphere and its hierarchical organization. (a) Sketch view in cross section showing circulation of water and earth; (b) areas of the earth showing the hierarchical converging of emergy to the mountain centers.

there is a similar distribution in gap sizes with many small gaps and fewer large gaps. If you walk along a line through the forest passing through gaps of the various sizes, the ups and downs can be thought of as oscillation over space, the spatial equivalent of pulsing in time.

ENVIRONMENTAL PATTERNS

Global Hierarchy

The three main outside sources of energy for the Earth's geobiosphere (Figure 6.1) are: (1) direct sunlight, (2) tidal energy transferred to the sea by the pull of gravity of the sun and moon, and (3) geologic processes caused by the heat emerging from deep in the Earth. These energy inflows are broadly distributed over the Earth. The processes of the atmosphere, oceans, and earth, as they interact with each other, operate a global system in which the ocean helps form land, and land develops mountains (Figure 7.3). As shown, solar energy is absorbed over the globe, especially by the seas, where it becomes heat and evaporates water. The solar heat and water vapor make winds, storms, ocean currents, and rains, especially over land. There rains and snowfalls produce rivers and glaciers. The large-scale, slow cycle of the land is driven by the rivers flowing down mountains, and the earth heat causing land movements from below. The earth processes converge rocks to form the center of the geological hierarchy, the mountains (Figure 7.3).

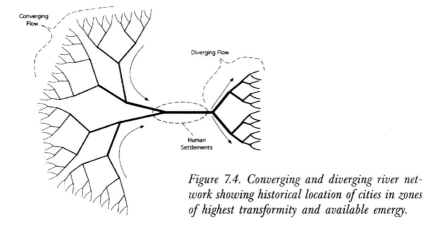

Figure 7.4. Converging and diverging river network showing historical location of cities in zones of highest transformity and available emergy.

This earth series is an energy hierarchy chain like the one moving from left to right in Figure 7.2. Many rapid processes in the sea support fewer large processes that turn over slowly in the mountains. The transformities and importance to humans increase in the order of this transformation series (Figure 7.3, from left to right).

Emergy Concentrations

In the places on Earth where the natural processes of wind, water, and earth converge, the transformities and concentrations of emergy increase. *Empower density* is a term for the rate of emergy use per area.[2] In the past, conditions in these places of empower concentration made the best places for human societies to concentrate because high-quality resources were converging there. Human cities often developed where streams and rivers converge and before the waters diverge again into their deltas (Figure 7.4). As a result, structures and functions of high transformity developed there. Maps of empower density and transformity can guide policies for locating developments in the future. New developments should be placed in areas with similar indices.[3]

Rains and snows already have a moderately high transformity when they fall on the land, and they are concentrated (transformed) further as waters run together to form rivers or accumulate as ice flows. Much of the Earth's surface is organized by the streams and glaciers that sculpt the landscape (Figure 7.4). In some places the waters converge— for example, the many tributary streams that combine to form the Mississippi River at New Orleans. In other places the waters diverge, such as in the Mississippi or Nile Deltas or waters flowing down and out from conical volcanoes.

When streams converge, not only is water flow increased, but also valuable substances are transported, especially sediment, nutrients, and organic matter. Transformity is highest at the places where waters and their contents converge most, usually at the coast, and here the emergy of the whole watershed is available. Here navigation, development of fisheries, and use of water for urban development are easy. Little wonder that the great cities of the pre-industrial past tended to develop at the centers of the hydrologic hierarchy: Cairo and Alexandria on the Nile, Vienna on the Danube, New York on the Hudson, and Shanghai on the Yangtze.

In the lower part of the watershed the physical energy of running downhill spreads water out into flood plains, delta marshes, and agricultural lands, where it stimulates productivity. An important hypothesis is that geomorphologic self-organization of the landscape gets reinforced so that the energy of elevated waters (geopotential) can help maximize productive uses of fresh water downstream. The energy of low salt content is a property of freshwater that makes plants grow as they transpire the water. It is measured as chemical potential energy relative to saltwater. The energy of the water running downhill makes floods that the wetlands need. Any plans to divert upstream river waters must also take into account the values generated by the lower floodplains and deltas that may be lost. Damming the Nile River, for example, caused loss of land and fertility in the delta.

Another place where the planetary system's energy flows converge is at the seacoast, where the interaction of water waves and sediments form beaches. Lagoons and estuaries behind the beach help to convert tidal energy into fisheries. Economic development along coastal strips is an example of developments attracted to areas of high empower.

Some energy transformation series are laid out spatially, their hierarchical branches easily recognized, like the river system in Figure 7.4. More often the components of an energy transformation hierarchy are together in the same area. Viewed from above, the system seems complex with many small items and large items together. See, for example, the landscape sketch in Figure 7.5c. However, the systems diagram (Figure 7.5b) separates the hierarchy into units in order of decreasing energy flow and increasing transformity from left to right. The center with concentrated information is on the right.

HUMAN SETTLEMENTS

The spatial and hierarchical organization of human settlements in agrarian landscapes was recognized early on.[4] Scattered rural people supported a village and received trade and services in return. Villages supported towns, and towns supported cities. Large, low-transformity energy flows of the rural system were transformed into higher trans-

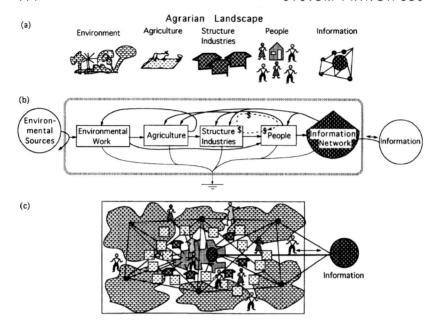

Figure 7.5. Model of the past agrarian landscape mostly running on renewable energies. (a) Sketch of five zones from the rural environment to the center; (b) energy systems diagram; (c) aerial view.

formity inputs to villages, whose products were further transformed into even higher level inputs to towns and cities. Transportation and utilities were effective when organized hierarchically, converging toward the centers of higher transformity. People and products were circulated into these centers along converging roads and then returned, often diverging again along the same corridors.

The hierarchical distribution of human settlements and the economy on the landscape is summarized with Figure 7.6. People and information are more concentrated in the centers where transformities are higher. More money circulates in the centers than in the rural areas. The emergy-money ratio is highest in the rural areas where money buys more real wealth, since many environmental resources can be used directly there without circulating much money. In the city centers, nearly everything for people's needs has to be brought in and purchased. Money buys less there because a higher proportion of the purchased wealth consists of human services for which payment is required. Knowledge, information, and financial transaction are concentrated in the centers. Money, materials, and information converge as they circulate in and diverge as they circulate out.

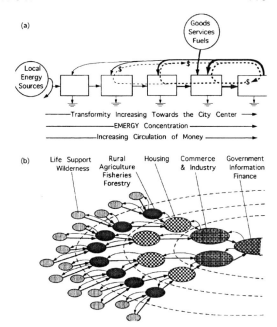

Figure 7.6. View of the convergence of properties to urban centers showing hierarchical concentration of emergy, population, materials, and circulation of money.

Environmental Zones Around Cities

Even in the landscape of the agrarian economy two centuries ago, there were zones of increasing concentration and transformity that a traveler would pass in going from wilderness inward to small cities. For simplified policy thinking, the zones can be aggregated into a chain of blocks, each representing a zone around the city center. For example, Figure 7.6 has five zonal blocks: natural areas, rural production, housing, industry-transport, and the center of information and finance (including business, banks, libraries, and government). Concentrations of emergy use, circulation of money, and transformity increase toward the center.

For an agrarian landscape of the past, most of the high concentrations of emergy and transformity originated from the broadly distributed, renewable environmental resources being converged into the city. The sketch in Figure 7.5 represents the agrarian economy that may come again when the global cycle is at a lower energy stage (see Chapters 5 and 12). Shared information has processing centers, but its territory is the whole landscape from which information is drawn and fed back. With larger territory and emergy content, the turnover time and time between pulses of shared information is longer than for the smaller scale units in the zones further out (at lower level in energy hierarchy). In many agrarian areas, such as Sweden in 1650,[5]

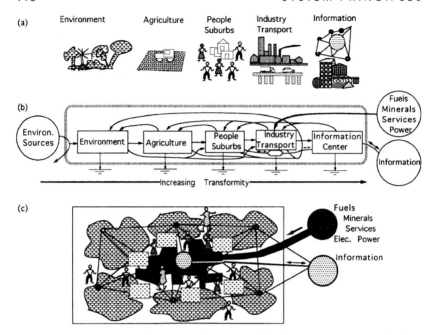

Figure 7.7. Model of the fuel-based landscape of urban America with suburban populations outside of the fuel-using transport-industrial areas. (a) Sketch of five zones; (b) energy systems diagram of the zones from the rural environment to the center; (c) aerial view.

rural exchange was mostly by barter, whereas silver coinage was used within the cities. Human settlements that self-organized over centuries functioned well with the natural pattern of energy hierarchy. Good policy in restoring ailing cities may come from re-establishing the spatial hierarchical patterns that maximize empower (see Chapter 12).

Regional Transformity Matching

As explained in Chapter 4, maximum empower requires a network of reinforcing feedbacks (Figure 4.2). Each flow must interact in a multiplicative way with another resource of different transformity either an order of magnitude higher or lower. In this way each pathway amplifies or is amplified. Figures 4.4 and 6.7 show how autocatalytic feedback intersections look when diagrammed. Reinforcing interactions link the zones of the city (Figure 7.6). For example, humans feed back their services to control and facilitate agriculture. Information from the city center feeds back from the central city to control and facilitate industry. We suspect that the interactions between levels in the energy hierarchy are most effective between adjacent levels. Information in organizations tends to cascade down one level at a time. The higher its transformity, the more levels something can influence.

GEOGRAPHIC REORGANIZATION WITH FOSSIL FUELS

Starting in the nineteenth century, the pattern of spatial organization was profoundly changed by technology and the availability of rich fossil fuels and minerals for development. Because these resources form slowly, they are essentially nonrenewable for the time scale of economic development. Growth was accelerated by a competitive race to develop these fuels, and the geographic pattern of economic activity was no longer restricted to those areas where earth energies could converge in a series of transformations. Instead of high transformities of human society coming up from the renewable earth energies, development could take place anywhere with access to the fuels and minerals, providing resources could be found to interact. For example, cities develop on fossil fuel providing they also have suitable lands and waters. Figure 7.7 sketches the patterns of life in the fuel-driven hierarchy for comparison with the sketch of the environment-driven agrarian landscape in Figure 7.5.

Fuels were carried directly to the cities with ships and pipelines. Because of the high net emergy yields, growth and developments were far beyond what had been possible with the original regional resources. Political power originally had a rural base where land captured the principal energy source, which was solar. Political power followed the empower concentrations to the cities. Empower densities became very high. High levels of people and information in the city put a strain on the environmental resources of air, land, and water. Uncycled waste accumulations caused serious pollution. Streams of air and water in and out were required for industry, housing, and waste dispersal. Fuels and derived products were sent out for use over the surrounding landscape. Fuel-based transport extended the territory of city-rural interaction and allowed suburbs to develop with a commuting culture and auto congestion. In Chapter 10 we will discuss the present state of cities and the policies they will need when fuel availability decreases.

Gradient of Investment Ratio

Fuel-based empower is most concentrated in the city centers, diminishing outward to the rural environs where the free environmental inputs are more important. The investment ratio (emergy of purchased inputs to emergy of local free inputs) introduced in Chapter 6 (Figure 6.6) measures this gradient. The ratio is high in the center of cities and low in the lower-energy rural areas outside. Maps of investment ratios can be useful in finding appropriate areas to site economic activities of

different intensities. For example, you might expect to make money by putting a copying business with a high investment ratio in the city center rather than in a rural area with a low ratio. Industries requiring more environmental inputs belong further out.

Electric Power and Information

Duplicating and processing information through the centers of television and computer networks is based on electric power. Electric power is required to support city people, information in computers, educated professionals, business transactors, government leaders, university researchers, doctors and nurses, artists, and television celebrities. Electricity has a moderately high transformity, intermediate between that of the fossil fuels and information. It has great flexibility in its uses and has become the main method of supporting higher levels of human society, especially those concerned with information. With fuels cheap and transportable, electric power plants to support cities and information have been installed everywhere.

All over the Earth in the twentieth century, reorganization of economic society has generated a spatial urban design running at its center on electric power and information. In Figure 7.7, note the typical zonation of the energy hierarchy around the modern, high information center. Industry and information are shown at the top of the hierarchy (to the right in Figure 7.7a).

Spatially, people and the information they process are concentrated in the urban centers, but the whole landscape is affected by this information. For example, though television and movies are produced in cities, the whole world uses these products. Information that is shared

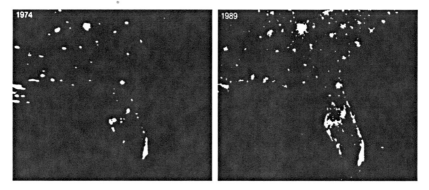

Figure 7.8. Night lights of Florida from satellite (Fernald and Purdum 1992): (a) 1974; (b) 1989. USAF Defense Meteorological Satellite Program, National Snow and Ice Data Center.

by many people has a large territory and the highest of all transformities on Earth. In those countries that are at the hierarchical center of the world system, people and technology are increasingly reorganizing around centers of information processing and storage. Part of the learned information of human society is becoming globally shared. However, if more centers are developed than can be supported by the energy hierarchy, some will collapse.

The new surge of global information sharing for the people in cities includes vivid views of global ecosystems and their devastation. Public support for protection of environmental systems is increasing. Urban public opinion is forcing resource managers to restore the life support system. See following chapters.

Night Lights

A very dramatic view of the spatial hierarchy of our high-energy world comes from satellite photos of the Earth at night (Figure 7.8). Each town, city, and metropolitan center glows with light. The pattern of smaller centers of town lights around larger areas of large cities shows the spatial patterns of energy hierarchy and concentrations of high transformity. Maps of the distribution of population are very similar to the night light photographs. Night lighting is one of the ways the urban civilization maximizes its functions by extending the hours when operations can continue. The photograph shows the way urban functions of our civilization are derived from electric power.

But electric power, with its transformity about four times that of fossil fuels, requires huge fuel flows and diversion of much of the energy of the mountain rivers that used to support other productivity such as fisheries and lowland soil renewal. With less fuel available in

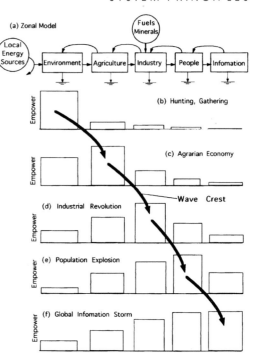

Figure 7.9. Zonal distribution of empower in five stages in the history of human development. Arrows show the movement of the wave crest of development. (a) Model of landscape hierarchy with five zones; (b) hunting and gathering peoples based on environmental systems; (c) agricultural basis for people before much use of fossil fuels; (d) predominance of industry and transportation using nonrenewable fuels and mineral reserves; (e) population surge based on increased empower; (f) current time of information production and use.

the future, renewable sources of electric power may control where the centers of information will continue.

Emergy Wave

Many urban centers are still growing in their levels of consumption, information, and lights, even though the availability of resources to support some centers is already beginning to decrease. As several popular writers have explained (see pp. 44–47), the development in the last two centuries has been a wave of ascendancy moving up the energy hierarchy. We can represent the wave by showing the relative empower in each stage (Figure 7.9). First there was a predominance of agrarian agriculture, then use of fuels and minerals for the industrial revolution, next a population explosion, then a concentration of emergy in the cities, and finally highest emergy in the worldwide sharing of information. The wave may be expected to generate more population and information than it can support. As the emergy flows of the fuels and their matching resources falter, we expect the climax to turn down, either crashing or descending prosperously depending on how well the world shares common purpose. Referring to Figure 7.9, how do we go from the pattern in graph (f) to a new kind of future consistent with the limitations of graph (c)?

SUMMARY

Energy hierarchies are organized spatially with materials and energy converging from dispersed producers toward consumer centers, with materials and emergy diverging back to the supporting territory outside. Using the rivers, glaciers, and deep earth heat, the geologic processes organized the planetary structures and functions spatially, from the broad oceans to the lands to mountains at the hierarchical centers. Coupled to the landscape, the human economy is organized with cities at the hierarchical centers, drawing their support from rural environmental production much increased by inflows of fuels and fuel-based goods and services. Centers have high concentrations of emergy use, high empower density, investment ratio, transformity, electric power use, money circulation, and information. From the centers the system sends out recycled materials, reinforcing services, controls, and pulsing impacts. Emergy indices are useful in planning to determine where in space activities may be compatible.

8
POPULATION AND WEALTH

Much of what happens in the growth cycle of a civilization depends on the human population. The number of people affects the economic productivity and vice versa. This chapter uses models to consider population, real wealth, and quality of life during the transition and turndown ahead.

FACTORS AFFECTING POPULATION

Populations depend on the balance of birth rates and death rates, which demographers relate to different age groups in order to extrapolate future trends. Authors cited in Chapters 2 and 3 indicate that global populations will increase well into this new century, but at a decreasing rate. Late in the twentieth century, lower birth rates in developed countries have been more than balanced by immigration from the areas of higher birth rates. During growth periods, births are stimulated by national policies and a global image of expansion and progress.

However, for the times of major change ahead, birth and death rates are likely to change sharply as society adapts to different priorities and availability of resources. Where resources for public health and medicine decrease, death rates increase. Where great social disorder causes loss of controls, there can be runaway births and deaths–observable in Africa–somewhat analogous to the loss of control of cell reproduction that causes cancer.

However, in an ordered society, depressed economic conditions reduce birth rates as people delay marrying and having children. A reversal in humanity's present global image of economic growth could change the social attitudes and reproductive behavior of families. In the past the more stable human populations developed cultural and religious customs to limit populations at levels that did not become pathologically crowded or undermine their resource basis. In 1999, the religious theocracy in Iran, responding to unemployment and urban crowding, announced a policy of limiting births.

Pulsing population oscillations are normal in many animal populations and were common in human history. The idea expressed in Chapter 6 is that pulsing oscillation is, in the long run, the most sustainable regime. Available resources allow a population to grow until it becomes crowded and drains those resources. Then because of reduced resources and negative factors of crowding, the population declines until resources accumulate again so that the cycle can repeat.

Wealth and Biotic Potential

The factors affecting population can be summarized with a model that relates population to resource wealth (Figure 8.1). The phrase *biotic potential* refers to the tendency for people to make more people and is shown as an *autocatalytic loop* in the diagram. Where there are no resource limits, the biotic potential can be expressed as explosive exponential growth, such as what was observed when rabbits were first introduced into Australia.

But the model shows how a population also depends on the real wealth of its economic assets. On the left side in Figure 8.1, the economic assets are shown interacting with the population to grow more people. In this model, reproduction is made proportional to the number of people and to the assets per person.[1]

Mortality is shown on the right side of the model in Figure 8.1. High concentrations of people cause death rates to increase, sometimes with disease epidemics. One pathway in the diagram is for normal mortality and one is for epidemics, both reduced by public health measures and health care in proportion to economic assets. Having related the economic condition to the population, next consider the reverse, the effect of the population on the economy.

Figure 8.1. Model showing factors affecting population numbers. The hexagon shaped symbol is used for consumer units. The lower pathway from economic assets represents the reduction of mortality and epidemics by health care and prevention.

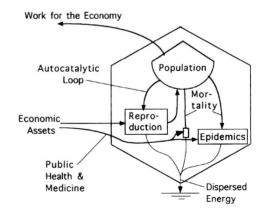

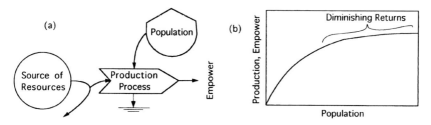

Figure 8.2. Production of economic assets by the population from using resources. (a) Energy systems diagram of the production process; (b) empower output as population increases.

POPULATION EFFECT ON THE ECONOMY

During the last two centuries of economic expansion, population growth helped to accelerate economic productivity. Practices, policies, and religious instruction favored reproduction, and thus increased reproduction rates and reinforced economic growth because population was a limiting factor. But in the late twentieth century, other factors began to limit the economy. The interaction of two factors—limited resources and population—is shown in Figure 8.2a. Where essential resources are source limited, the useful economic output (*empower* in Figure 8.2a) has "diminishing returns" with population increase. The response to increased population (Figure 8.2b) comes from the textbook mathematics of limiting factors.[2] Production (empower) increases at a diminishing rate as the population increases and resources become limiting. Empower production levels fall, even though more people are added. Authors cited in Chapter 3 are already concerned about the shortage of meaningful jobs in many places.

Humans have become the world's information processors while deriving their support from the Earth's life support. In theory at least, the trend of self-organization should be toward the optimal population size that maximizes global empower.

Real Wealth per Person

If resources are limiting, further increase in people means decreasing wealth per person, an effect already observed in much of the world. Annual emergy use per person shows that in some countries—such as India, for example—the population density is already above that optimum for maximum empower (Table 6.2). Adding more people without more emergy means a lower standard of living. Paul Ehrlich's 1968 bestseller *The Population Bomb* updated the earlier warnings of Malthus. Potentials for disaster are borne out in Africa, where the systems of population control by tribal culture and chronic disease were

"You know, Vern ... the thought of what this
place is gonna look like in about a week just
gives me the creeps."

THE FAR SIDE by Gary Larson © 1988
FARWORKS, INC. Used by permission. All rights
reserved.

disabled by European colonialism and health care, without a suitable
substitute.

POPULATION CARRYING CAPACITY

In a large-scale experiment that gave insight about the whole Earth,
ecosystems and people were sealed together in Biosphere 2, a giant
glass chamber at Oracle, Arizona, for two years (Figure 8.3). *Biosphere*
2: Research Past and Present (1999), edited by Bruno Marino and H. T.
Odum, summarizes the scientific results.[3] This experiment in ecological engineering dramatized the difficulty of supporting eight people
on three acres in a closed system, even when unlimited electric power
was available.

Remarkable engineering controlled the interior microclimates to
form areas for desert, savanna, thorn forest, rainforest, mangroves,
and agriculture. The enclosure was started with soils rich in organic
matter, a small ocean, and several hundred plant species appropriate
to each area. Normal insects were not added because of federal restric-

Figure 8.3. View of Biosphere 2, the glass-enclosed living model of Earth in the mountains of Arizona. Photo by Scott McMullen, Biosphere 2 Center, Inc., Oracle, Arizona.

tions on importing species, and only a few insect species appeared from accidental seeding, not enough to support the birds added. The plants flourished, but without the usual means for pollination and seed dispersal. As waters percolated, soils began to form a typical profile.

The studies showed that several years are required for the photosynthetic production of the plants and the consumption by soils, microorganisms, and people to get in balance. While out of balance carbon dioxide (CO_2) accumulated and when some of it was bound into the carbonates of concrete, ocean water, and soils, oxygen (O_2) was bound with it (since O_2 is part of CO_2). Since lowered oxygen was a human hazard, a tank truck was brought in to replace the lost oxygen.

This great experiment dramatized the way a symbiotic relationship is required between the life-support ecosystems and human population. The difficulty with getting a life-support system going and its enormous cost helps us appreciate the Earth's life-support system and how essential its integrity is for civilization.

As many authors have written, closed systems like Biosphere 2 and our *Spaceship Earth* have limits to their ability to support population. In Biosphere 2, by working hard in the farming areas and losing weight, the eight people inside were just able to support themselves for the two years of closure. The carrying capacity was between six and eight people.

Many authors seeking a steady state for human populations on Earth (see Chapter 3) discussed the concept of *carrying capacity*. As first used for wildlife, carrying capacity is the size of a species population that could be sustained per area of supporting ecosystem. But the carrying capacity for humans is the number that can be supported on the local area *plus* the resources that can be brought into the area from elsewhere, usually by purchase. Many authors (see Chapter 3) now use the term *ecological footprint* for the local area plus the outside area necessary to supply the imported resources.

But it is the money generated from the resources in a local area that determines how much is purchased and imported from outside. We explained in Chapter 6 how a region's *investment ratio* (purchased empower divided by local free empower) indicates its potential for development and thus its carrying capacity. As global resources decline, the investment ratio, carrying capacity, and footprint also decline.

THE SYSTEM OF POPULATION AND ECONOMY

To dramatize simply the future possibilities for global population, we combined the factors affecting population (Figure 8.1) with the effect of the population on economic productivity (Figure 8.2). The result was the model in Figure 8.4a.[4] The model showed inputs of renewable energy and nonrenewable reserves, and population interacting to produce economic assets. Population was represented by the hexagonal consumer symbol (Figure 8.1).

Computer simulation of the model showed how economic assets and population could interact in the future as the nonrenewable resources are used up. In the simulation represented by Figure 8.4b, the population growth was at first autocatalytically accelerated, thus maximizing the rate of resource use and production in the growth period. During the growth time represented by the simulation, political and religious policies encouraged population reproduction (as do, for example, Catholic, Mormon, Islamic, and Chinese doctrines).

Then the simulation showed economic assets beginning to level while the population was still increasing, as in our current time. With less health care per person, disease mortality increased. This period may be like the start of the new century, when we observed old diseases increasing again, and new ones like AIDS, drug-resistant malaria, and new viruses straining the public health systems of many countries.

The simulation next showed population increasing beyond the time of maximum economic assets, so that the standard of living (emergy per person) declined rapidly. The bracket in Figure 8.4b marks the critical time for world society discussed by many authors (see Chap-

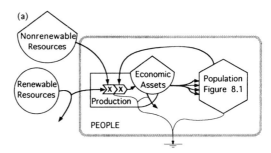

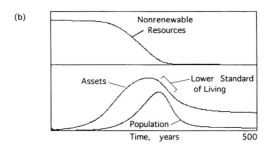

Figure 8.4. Simulation model of world economic assets and population (People). (a) Energy systems diagram of main features; (b) typical result of simulating the program People *in BASIC.*

ter 3). At that stage there were more people than necessary for the maximum possible production. The simulated death rates were high and birth rates fell as the real assets per person decreased. Already there are policies to reduce births in China and Iran.

The simulation showed that there could be a very sharp turndown from rapid population growth to rapid decline. The increased infant mortality in disturbed countries and cities showed what happens when the energy for public health and medicine per person is small. The policy challenge is to reduce the population first so as to keep the annual emergy use per person constant during descent.

Genetic Viability

To sustain healthy genetic stocks requires that more individuals be born than reproduce so that there can be selection against the genes of birth defects and disease susceptibility. In the past there was more selection during times of greater infant mortality. Finding a more humane method of maintaining genetic viability is a challenge for the future.

DISTRIBUTION OF WEALTH

With worldwide television, people everywhere are sensitized to the uneven distribution of real wealth between and within countries.

International Distribution of Real Wealth

Authors cited in Chapter 3 described vividly the increasing gap in wealth between developed countries that use much of the world's resources and many other countries with large populations that have their real wealth exported. Emergy use per person (Table 6.2) is a good index, because it includes both inputs obtained free from nature and those purchased with money. Some developed countries have fifty times more emergy use per person than some overpopulated nations. Many authors cited in Chapter 3 described eloquently the huge waste in developed countries. In other countries there is not enough support per person to sustain their productivity. Maximum empower theory predicts that if an excess of poorly used people is not maximizing global productivity, major reorganization will occur.

Hierarchical Distribution of Wealth

We explained in Chapter 4 how workable systems have units at each level in the energy hierarchy chain supported by larger energy flows from below. In Figures 4.5, 4.8, and 5.6, flows and quantities decrease as the scale of size and transformity increases. In other words, there is a natural pattern for distributing wealth between the bottom and the top of the hierarchy. A similar distribution of real wealth and/or income may be expected for a smoothly operating human economy. Policies that cause dysfunctional curves of distribution (too many rich or too many poor) may be energetically unsustainable, which eventually makes them politically unsustainable as people change public opinion to fit need (see also Figure 12.2).

SUMMARY

World populations and the Earth's environmental systems are reorganizing their relationships during the temporary surge of nonrenewable resources of our time. Biosphere 2 helps explain the concept of carrying capacity and the need for global symbiosis. The Earth's carrying capacity for people depends on the real wealth produced by the geobiosphere's environmental systems plus wealth based on the use of fossil fuels. For a local area, the emergy investment ratio of the region indicates the emergy that an area can purchase, but that ratio (and its implied carrying capacity) will decline with less fuel use.

Models were used to explain the dependence of population on economic assets and the diminished economic wealth as population increases. Global population was related to resource use by simulating a model of world population growth and decline. The run showed lower standards of living during transition and descent and the possibilities of a sharp turndown. World population may already be exceeding the optimum level for maximum productivity of the system of humanity and nature. The standard of living measured by the annual emergy use per person may have reached its maximum. The energy hierarchy concept suggests a functional basis for the distribution of economic wealth and power among people.

III
POLICIES FOR TRANSITION AND DESCENT

In Part III, we use energy systems principles from Part II to understand trends and suggest policies for the future. Chapters 9–12 offer policies for the period of climax and transition, and Chapters 13–19 for the later time of descent. Policies based on understanding could be the difference between a soft landing and a crash. Let's start with the global scale.

9

THE GLOBAL NETWORK

The international pattern of humanity and nature is now distorted and unstable with uneven surges of population, information, economic exploitation, environmental destruction, and military threats. To hold the world steady now while preparing for a prosperous descent ahead, we need to stabilize capitalism, protect the Earth's production of real wealth, and develop equity among nations. This chapter proposes policies to improve international relationships and world empower. Here, we wear an international hat, seeking what is best for the whole Earth.

The Expanding Scale

Starting with the industrial revolution, the accelerating use of nonrenewable resources allowed larger and larger areas to come under common organization. Struggles among city-states in the sixteenth century became struggles of small nations in the seventeenth century, struggles of alliances in the eighteenth century, struggles of empires in the nineteenth century, and struggles between two giant areas of influence (the East and West) in the late twentieth century. Now the world struggles with the forces making society into a single system.

More people are thinking globally. Television programs show the world systems of weather and business. Millions use e-mail and the global Internet. The world economy has become a single system, its capitalism driven by multinational corporations. Jobs in individual countries now depend on economic trends worldwide. International exchanges control smaller-scale developments. Global climax, like the climax described for smaller ecosystems (see Chapter 5), is increasing efficiency by evolving cooperation.

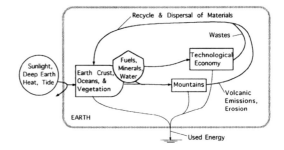

Figure 9.1. Diagram of the Earth system, its bio-geochemical cycle, and the role of the techno-logical economy.

GLOBAL ENVIRONMENTAL STRESS

The problems of pollution are put in perspective if they are seen as a disturbed part of the Earth's material cycles (Figure 9.1). Without humans, the normal biogeological system (earth crust, oceans, and vegetation) accumulated stores of rock, mineral deposits, fuels, and water. These concentrate in mountains in surges from which volcanic eruptions and river runoffs dispersed and recycled materials. In Figure 9.1 the lower loop (through the mountains) was the main pathway before economic development.

The intensive consumption of fuels and production of wastes by the modern economy speeds up the Earth's geological processes. The human technological economy on the upper loop in Figure 9.1 is a competitor in parallel with the original recycle processes. Civilization has accelerated use of mineral accumulations, dispersed mineral deposits, and released wastes not unlike the original geological cycles. The acid wastes from our smokestacks are chemically similar to the emissions of volcanoes. The problem is having too much and in the wrong places.

Worldwide interest in protecting the global environment is evolving, although not everyone sees the connection between economic activity and environmental degradation. There is worldwide concern about our shared atmosphere, its increasing turbidity, carbon dioxide, disrupted ozone dynamics, acid rains, and the greenhouse effect changing the climate. More and more we see transfers of one country's waste outputs into air and water affecting other countries. Toxic substances with high transformities from industrial nations are forcing the ecosystems of surrounding nations to provide treatment. Even solid wastes are now being transferred from one country to another by businesses that operate across borders in order to maximize profit.

Market evaluations of environmental stresses are much too low (see Chapter 6). By calculating emergy values of international waste exchanges, the extent of the impact can be put on a common basis with economic values. For example, the emdollar value of the acid sulfur gases discharged by the smokestacks in the United States can be compared with the dollars in the cleanup budget of Canada where the impact is felt.

Much of the Earth's air purification is done by processes in the atmosphere, oceans, and vegetation. However, most of the world's forests are being rapidly stripped by capitalistic investing. Consequently, less green work is available for life support. Global empower is not being maximized when economic markets reduce the Earth's contributions. According to the principle of transformity matching (see Chapter 6), the high-transformity goods and services of the fossil fuel–based economy need to interact more symbiotically with environmental resources of lower transformity. Measures for world forest restoration need to take higher priority (see Chapter 16).

GLOBAL INFORMATION AND THE ECONOMY

Everyone feels the new global excitement of international information sharing through television, the Internet, multinational finance, and exchanges of people. Because information shared widely has a higher transformity than most of the economic system (see Chapter 4), information can control the economy, more than vice versa. Practical experience in software marketing already recognizes that you get money for information by giving it away first to get it established. Then you can be paid for services, advertising, and improvements to keep it going.

Yet efforts by information industries in developed countries to restrict information, demand money for information use, and increase profits in relation to patents and copyrights, inhibit global empower and can't succeed long. Money needs to be coupled to information production in a less-inhibiting way, enough to pay costs of developments, but not enough to limit use. Because of the high transformity of information (including art, literature, and technology), it is so flexible and transportable that it really can't be regulated well, nor should people try.

WORLD ORGANIZATION

The global sharing of information is changing in the international order and politics of governments. As shown in Chapter 3, interpretation, attitudes, and predictions of future consequences are widely different. Some fear homogenizing as loss of structure that would lead to the anarchy and the decentralization of the Middle Ages. Others, in a policy analogous to that of the ancient Roman civilization, call for isolation and walls to protect the western civilization against homogenization of peoples and cultures or barbarism. Many fear the loss of national unity and character because of international capitalistic control of governments. Some imagine the replacement of the present pattern of nations and states with a decentralized network of people connected by communication links everywhere similar (Figure 9.2a). For each futuristic idea, some react with denial, some with fear, some with militancy, and others with optimism.

Figure 9.2. Communication network. (a) Initial homogeneous pattern; (b) emergence of hierarchy in self-organization.

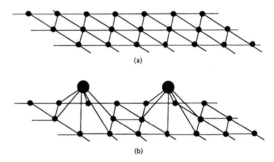

Prediction with the Hierarchy Principle

We can distinguish between and rationalize conflicting views with the energy hierarchy principle (see Chapter 4). Systems form energy hierarchies because this design maximizes performance, displacing the other alternatives (units homogenized, in anarchy, or in isolation). For example, human populations distributed over an area automatically form villages, and some of these rise to become the regional centers for the villages. Thus there is no danger that the hierarchy of human settlements (villages, centers, states, nations, etc.) will disappear.

The idea offered by some authors that the world can become a homogeneous decentralized network of information processing nodes is sketched in Figure 9.2a. However, it can be expected to self-organize into a hierarchical network suggested by Figure 9.2b.

For the relationships between political units, sharing information (common beliefs) appears to be more energetically efficient than organizing power spatially with military defenses. Already, something of a new global culture is developing with the power to eliminate wars by interconnected futures and shared aspirations and attitudes for peace.

But global sharing of information takes substantial resources to develop. It may be restricted to those ideals that are important at the global level. At times, important information seems to be drowned out on television by the huge volumes of advertising, entertainment, and unselected conversation. How much the hours of television viewing each day are contributing to organization and productivity remains to be evaluated. Examples of important messages that need to be shared globally are: protecting the purity of the global atmosphere; maintaining cordiality and trade between neighboring countries that are culturally different; and sharing technologies that are useful anywhere.

It has been called a paradox that there is a spread of global information and economics and at the same time an intensification of separate efforts by local groups to hold on to their special languages, heritages, cultures, arts, and religions. This is no paradox, just properties of a developing hierarchy in which people can be effective by being different about what is small scale but united about what is large scale.

In 2000, many isolated populations were still out of range of the international media, with an inclination to conquer others rather than share information. Large nations have exerted collective policies to force small warring parties to reprogram their ideals. For example, military force was used in the 1990s by the North Atlantic Treaty Organization (NATO) to force *peace thinking* in the Balkans. Is this a trend?

The Symbiosis of Information Sharing

We have a neat computer simulation of a mini-model with two competing units that share information.[1] It shows dramatically how a common pool of information used by each maximizes the total productivity of both participants (maximum empower), causing coexistence and a special kind of symbiosis. The implication is that information needs to be shared and unrestricted. It would help if democratic, one-world ideals could be consolidated by spreading knowledge and discourse more freely into all the corners of world society. For this, cheap electric power is essential.

In 1980s Iran, the Ayatollah Khomeini organized a revolution by distributing tape recorders and tapes so as to create shared information about a doctrine in minds hungry for guidance. Distributing a different message of shared belief in peaceful relations among nations might be achieved by an international giveaway of television sets and VCRs in place of troops. Fortunately the spread of common thinking through global television is already well established in most places and increasing incredibly quickly. To sustain peace in times of cuttingback, priority needs to go toward sustaining this global means of communication sharing.

Global Pluralism and Migration

The mixing of races, languages, and former national people in this century is generating a global mix rich in genetic and cultural information. Some countries like Australia and the United States are "melting pots," developing more tolerant attitudes, laws, and institutions that appear to increase empower and economic vitality. Ethnicity and pluralism are the new buzzwords. The economic and peace-sustaining advantages of a unified culture are contagious, helping isolated groups reduce prejudices.

More people are moving across borders. Although some of the pressure is from revolutions and political oppression, much of it is from overpopulation. Increasing numbers of people no longer means much increase in production (Figure 8.2), while the emergy per person decreases. The trend in developed countries is to limit immigration because they do not want to decrease their standard of living or increase unemployment. For example, restrictions on immigration were added by some developed nations in the 1990s. Bermuda, an island with only a little land, required large payments of money for immigrant citizenship.

Net emergy of immigrants, and of cultural and scientific exchanges, can be evaluated as part of the balance between nations. Immigrants bring the emergy value of themselves and their education to be added to the host country's total. In each country immigrants are transported, housed, and fed, but if employed they contribute to the economy. Some send earnings out of the country. There are welfare costs where immigrants are not employed. The various pluses and minuses of the migration of people can be considered in the total calculation of emergy value gained or lost between countries. International policies on immigration and exchange are best set after emergy accounting has established what policies would be of net benefit to the world as a whole.

The emergy per person for many countries in the recent past is given in Table 6.2 and the national emergy balances in Table 9.1. The wealthier nations increased emergy per person by importing valuable raw resources. Then population migrated toward those nations with the higher per person values. In other words, the people of nations that exported their resources had to migrate to developed countries to use their own resources. A better way for these nations to increase emergy per person is to reduce their populations and use their own wealth instead of exporting it.

EMERGY AND EQUITABLE EXCHANGE

Nations exchange emergy with their trade, finance, international aid programs, population exchange, military actions, and information. One of the principles (see Chapter 6) suggests that the global system as a whole maximizes its performance when the exchanges are equitable. Then each nation can contribute at its potential best and is not drained by another.

According to economic concepts, trade can increase overall productivity if each area supplies what it has that is special. Such specializa-

Table 9.1—Balance of Traded Wealth Evaluated with Emergy*

Nation	Emergy from Within %	Emergy Received / Emergy Exported
Netherlands	23	4.3
West Germany	10	4.2
Japan	31	4.2
Switzerland	19	3.2
Spain	24	2.3
U.S.A.	77	2.2
Taiwan	24	1.89
India	88	1.45
Brazil	91	0.98
Dominica	69	0.84
New Zealand	60	0.76
Poland	66	0.65
Australia	92	0.39
China	98	0.28
Former Soviet Union	97	0.23
Liberia	92	0.151
Ecuador	94	0.119

*Different years in the 1980s (Odum 1996, Table 11.1).

tion helps join two areas in peaceful cooperation. However, to make sense, the emergy benefits from the trade need to exceed the emergy required for the arrangements and transportation. Some local capacity for self-sufficiency needs to be retained to avoid the risks of disruption if something interrupts global exchange. To be beneficial, the exchanges of real wealth must be equivalent on an emergy basis. Emergy evaluation shows that exchanges are often not equal.

Trade Inequity

In the trade between nations, it is often assumed that exchanges based on market prices are fair. From Chapter 6 we know that money does not measure real wealth because it represents only the work of humans involved in developing the products. The work by the environment is not paid for and thus is not represented in the price. Market prices underestimate the real wealth value of "raw" commodities. In international trade, the raw commodities contribute to the buyer's economy more than is exchanged for them in return (Table 6.3).

As a consequence, the trade inequity between underdeveloped nations supplying raw products and the developed nations buying the products is huge. Table 9.1 shows the great difference in trade equity

From *EARTHTOONS*
by Stan Eales. Copyright
© 1991 by Stan Eales.
By permission of Warner
Books Inc.

between the rich nations with positive emergy trade balances (ratios greater than 1 to 1) and the poor nations with negative emergy exchanges (ratios less than 1 to 1). Charles Hall also found that externally financed developments were crippling developing nations.[2]

Even labor-intensive products may carry more emergy than is in the money received. The more rural a nation is, the more of its inputs go directly from environmental resource to human support without the circulation of money and intermediate human handling. In rural nations people can have subsistence crops, catch their own fish, and cut their own wood. They can charge less for their labor because they are supported partly by unpaid environmental inputs. The services that they sell include the hidden emergy contributions of the free commodities they use. A dollar of rural service represents more emergy than a dollar of urban service.

Developed countries like the United States may be on the short end of some emergy exchanges. For example, Figure 9.3 shows one dollar circulating between Japan and the United States in the early 1990s. Japan received an average 7.6 x 10^{11} solar emcalories in U.S. wealth (three emdollars[3]), whereas the United States received only 1.7 x 10^{11} solar emcalories (one emdollar) from Japan. The money traded was equal, but Japan received 4.5 times as much real wealth as measured

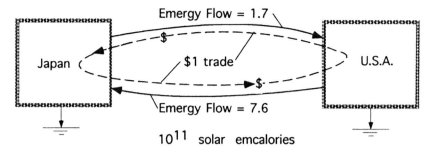

Figure 9.3. Uneven exchange of real wealth (emergy) between Japan and the United States shown by circulating one international dollar. Emergy-money ratios are listed in Table 6.1.

with emergy or emdollars. The emdollar imbalance when Alaska sold salmon to Japan was nearly 10 to 1.

Nations watch their overall balance of trade with other countries in dollars, since an imbalance accumulates money on one end, affecting the currency exchange rate and international debt when dollars come back to purchase bonds and property. Trade balances should also be evaluated with emergy (emdollars). As shown with Figure 9.3, the different emergy-money ratios affect the conclusion about the overall benefit.

National Export Policy

Where policies favor the export of raw commodities such as minerals, fuels, forest products, and agricultural products, a nation's wealth is drained. In historical colonial days the underdeveloped areas knew something was wrong, because the developed centers (the colonizing country) got wealthy and luxurious while the areas supplying the raw products remained dirt-poor and backward. Efforts were made by the rich centers to send aid, but there was no scale to tell how much was fair.

In the twentieth century, students from underdeveloped countries went to developed areas for training in economics, and were taught the supposed virtues of the present trade system. They went back home and inadvertently made their system worse by encouraging their countries to sell cash crops that export more emergy than received. After the exporting of resources started, the government and the development managers were "hooked"–accustomed to cash money and its tax–even though only the government and a few in the exported business benefited. Free market policies intensify the trade inequities.

A better policy uses the raw products at home. When the emergy is used by the home economy (instead of being exported), the real wealth is increased so that more money can be circulated without inflation. Home use of resources accumulates capital faster than borrowing or letting outsiders develop the resource because home development doesn't send interest or profits abroad.

Keeping the real wealth of old growth forest woods of Oregon and Washington within the United States instead of selling them abroad increases the real wealth of the home states. Keeping the wood at home would lower prices, helping U.S. citizens to buy houses. The United States in these sales has been acting like a colony.

Furthermore, the clear-cutting of old growth forests is a loss of the natural capital necessary to sustain production. For their long-range benefit, the forest industry (managers, investors, lumberjacks, and saw-mill workers) could adopt a more sustainable cutting cycle now rather than after clear-cutting the last virgin reserves.

The industrial farm states of the United States export corn, wheat, rice, and soybeans to the world, and agriculturists take pride in feeding the world and getting foreign exchange money for the United States. While these exports carry much more real wealth out than is received, this inequity does return some equity to poorer nations.

Industrial Agriculture and Migration of Rural People to the Cities

A century ago most people of the world were on farms, raising much of their own food, fiber, fishery, and forest products, plus a little extra to sell for necessary goods. But with the industrial revolution, fossil fuels were used to increase the yields of agriculture, aquaculture, and plantation forests. Machines did most of the rural work and the people went to the cities to make the machines, but it was not planned.

The cheap food produced with fossil fuel–based technology and chemicals undercut millions of less-industrialized farmers, reducing the value of their services. Displaced from their land and its free emergy, they went to slums in cities, becoming an urban labor pool aiding low-wage industries. The disruption was terrible as families were forced into urban poverty without adequate preparation or understanding. In the United States the disruption of this process was at its worst in the 1930s and was one of the causes of the Great Depression. Urbanization is still underway in many countries of South America, Africa, and Asia that are importing food from countries with industrialized agriculture. The stripping of the forests, soils, and fish stocks by outside capital investment lowers rural opportunity even more.

Cheap food also encourages overpopulation and international dependency, putting futures at risk. A more diversified agriculture in developed nations, with more consumption at home, would give the farm areas more jobs, more real wealth, and allow other countries to stabilize their own agriculture.

Foreign Aid and International Loans

Foreign aid has been given by one country to another on a hap-hazard basis, and is often controversial. *Aid without strings* sends an emergy benefit. The world was made stronger after the Second World War by the programs of aid to the vanquished nations of Japan and Germany.

Since trade usually causes more real wealth to go to the developed country, emergy value could be sent back to the less-developed nation in some other form such as agricultural technologies or education appropriate for the circumstances. After calculating the emergy imbalance in trade, loans, migration, military protection, and information exchange, foreign aid could make up the difference.

So-called foreign aid in the form of loans is entirely different, a form of economic colonialism. When a less-developed country borrows from a developed country, the foreign money received represents less real wealth than its own money does when used to repay the debt and interest. The buying power of what is loaned is much less than the buying power of that which is repaid because of the disparity in emergy per dollar. Therefore there is a severe drain on the economy of the borrowing nation.

For example, Figure 9.4 shows the real wealth inequity in an international loan where the countries have different emergy-money ratios. The dollar that the United States loaned to Brazil in 1990 bought U.S. goods with an emergy value of 7.6×10^{11} solar emcalories. (See Table 6.1 for emergy-dollar ratios.) Later when Brazil paid back the dollar plus 10 percent interest, that money bought an emergy value of 22.2×10^{11} solar emcalories from Brazil.[4] Counting the interest, the United States gained three times more real value than Brazil (equivalent to paying 300 percent interest). To maintain equity, international loans should be made on an emergy basis. Without the absurd profits, there would be less overgrowth and waste in the developed countries and more productivity in the undeveloped nation. Many authors have made these points in public affairs books (see Chapter 3). The emergy measures show quantitatively how bad the "foreign aid" has been.

Repayments of loans made by U.S. banks are not only economically destructive to underdeveloped borrowers, but are potentially disastrous to U.S. lenders too because of defaults and increased stress to international financial markets. After default some loans have been refinanced to reduce the interest rate. For example, an overdue loan from Brazil was sold on the debt market for a quarter of its original value, which is closer to the correct repayment were the transaction calculated with emergy equity.

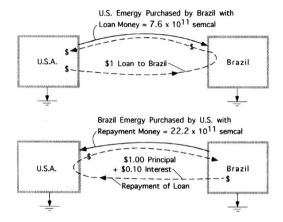

Figure 9.4. Unequal exchange of real wealth (emergy) for each dollar of international loan between the U.S.A. and Brazil. Semcal = solar emcalories. Emergy-money ratios from Table 6.1: Brazil, 2.0 x 10^{11} semcal/$; U.S.A., 7.6 x 10^{11} semcal/$.

Loan Policy

International loans should be recalculated and renegotiated with emergy accounting. It is quite possible that many underdeveloped countries have already overpaid their debts and could send an invoice to international banks for recompense.

In the interest of maximizing the real wealth of the world economy, global policy is needed to tame those aspects of international capitalism where the profit motive and lending, however well-meant, strips resources and inhibits the productivity of underdeveloped nations. The free trade agreements NAFTA (North American Free Trade Agreement) and GATT (General Alliance on Trade and Tariffs) approved in 1994 are not mutually beneficial to participating nations. Since free trade and finance calculated in dollars is unequal and unfair to undeveloped nations, they should require special trade treaties that maintain emergy equity.

World banks could propose a policy that loans have repayments and interest rates corrected for differences in the emergy values of currencies. Without emergy corrections, those responsible for the welfare of a nation should prevent borrowing from any country with a lower emergy-money ratio. This is a major reason for replacing free trade with equity-balancing trade treaties.

Tourism

When people have more money and resources than the minimum required for their bare necessities, they often travel to other areas as tourists. This interaction of an economic activity with nature is a kind of matching of high- and low-transformity inputs. Tourists bring and spend money used later by the host areas to purchase fuels, goods, and services. Many underdeveloped countries in the 1990s increased tourism to gain income. The international visits help educate the tourists of the world about the cultures, environments, and problems of our family of nations.

However, in many cases tourists used up or carried away more real wealth (emergy) than they contributed. More emergy of natural resources may be disrupted to build and supply hotels than is in the money the tourists bring in. Those promoting developments think that connecting the environmental resources to the economy is increasing wealth because they think that income measures wealth. Often these developments provide profit for a few while taking resources away from the local people. By causing local prices and taxes to rise and the buying power of currencies to decrease, tourist developments can displace native peoples from lands, fisheries, and livelihood. Some tourism is a luxury that does not reinforce world productivity.

However, low-density tourist development can be planned to add to the local system without diminishing it. Setting an upper limit to the investment ratio (introduced in Figure 6.6) can help optimize the intensity of developments For example, a resort plan for Nayarit, Mexico, made by a Cousteau project, recommended a low investment ratio.[5]

The supply of people wealthy enough to support tourism may have been at its maximum around the turn of the century at the crest of runaway capitalism and inequitable distribution of wealth. Later, travel costs are likely to be greater and fewer people will have the excess money and leisure time to travel. With declining resources, emergy-cheap communication can replace emergy-expensive travel. Tourist development boards in tourist towns, anticipating decrease, should have plans for different uses of facilities. For example, developers and architects might build into hotels the means to use them later as residences or civic centers for local people.

DEFENSE AND BALANCE OF EMPOWER

The traditional balance of power among the nations over the Earth has been organized by defenses around centers. Somewhat like the territorial defenses by birds during nesting, earlier human societies developed spatial organization according to the empower each could bring to its territorial border. A considerable part of earlier economies went into border skirmishes by which territories were contested. Military action or threat of action tested each center's ability to defend its area, so that each expanded its area of influence only as far as it could use its available resources competitively. In other words, some border wars were part of the process of organizing the global landscape into national units appropriate to the availability of emergy resources. Small

HAGAR THE HORRIBLE by Chris Browne. Reprinted with permission of King Features Syndicate.

nations obtained defense protection by joining defense alliances. Self-organization made divisions of the landscape that were equivalent in hierarchical position with similar empower. Such patterns insured a more even use of the resources of land, water, wind, and sun on homogeneous landscapes.

An area with a new innovation in agriculture, industry, military power, technology, or social institutions could assemble more empower and expand its area of influence for a time, incorporating the empower of other centers. The first nations developing fossil fuels were so powerful that invasions of the less-developed nations occurred on a global scale. As history shows for the nineteenth and twentieth centuries, rising energy encouraged competitive wars and made them terrible. Wars were won by high concentrations of empower that included high-transformity organizations and weapons. But after the new ways spread and were adapted elsewhere, areas became more equal again.

For two centuries of growth, the military forces of dominant countries forced other areas to accept profit making by private companies. Examples are Britain's support of the East India Company, the United States's dethroning of the queen in Hawaii for the benefit of sugar companies, and the use of marines to establish fruit companies in Central America and sugar cane corporations in Puerto Rico. Where political control was withdrawn, the *economic colonialism of emergy inequity* often continued.

The mechanism of military-aided capitalism and monopoly are analogous to the low-diversity weeds that maximize the empower early in ecological succession when there is an excess of unused resource. When expanding fuel use favored growth of dominant countries, the military-aided capitalism maximized empower of colonization by reinforcing the competition winners. The colonial invasions by the dominant countries were extensions of the fossil fuel civilization in its early growth stage.

At the crest of growth as resources decline, nations have fewer means to encroach on others. The availability of resources for any one country to dominate the world diminishes. Then it may become easier to develop a more even empower distribution around renewable sources. National boundaries can be readjusted to follow empower distributions.

Expenditures on defense decrease. Some countries formerly held together by military force may fragment. As the mechanism strengthens for holding world order with shared beliefs, the old ways of territorial defenses and alliances can decrease. Instead, shared military forces can control terrorism and groups that try using violence. Unfortunately, weapons of extremely high transformity such as the nuclear bombs are still in existence and might get used without the consideration of global safety.

Evaluating Military Power

Because emergy can summarize the different aspects of power, it can be used to evaluate national defense and the international balance of power among nations. For example, the net benefit of foreign trade of the United States in 1987 (Table 9.1) was made possible in part by an overseas defense force deployed to defend national interests. Because of the importance of fuels and strategic materials, emdollars involved in defense were about twice that of the dollars spent. Even so, emergy evaluation showed a benefit to the United States in net trade several times larger than the emergy used to support the overseas armed forces in a peaceful status.[6] On the other hand, emergy evaluation of Swedish power in 1650 showed little if any net benefit from its overseas army, which was frequently at war.[7]

If the results of emergy evaluation of military potential were recognized by all nations as accurate predictors of who would win a conflict, disputes could be settled by emergy computations rather than by war. If such calculations had been done before the Vietnam War, the ultimate cost could have been anticipated and the war avoided. Whereas the United States was using half of the world's fuel resources in 1946 at the end of the Second World War, by 1966 at the time of the Vietnam War, the U.S. share of world fuel consumption was down to a third. Its relative power to extend a defense umbrella toward Asia had been reduced. Judgments about war in 1966 were made by leaders whose attitudes were conditioned earlier when resource potentials of combatants were not so even.

Miscalculations on the ability to win have caused many wars in the past. Perhaps many conflicts can be settled peacefully by emergy calculations if the facts on the uneven capabilities for delivering military power can be spread by international television and believed. Emergy valuations of strategic power can prevent actual tests of military strength.

Wasteful arms races were caused by countries trying to gain security by having superior forces and technology. Arms races can be prevented with better knowledge of risk using emergy. The military power delivered at great distance is evaluated by the resource minus that used in transport and delivery. A reverse arms race (reduction) can occur if each country reduced its defense, keeping only enough emergy value in its military strength to defend its own border, but not enough to equal its opponents after transport to their area. Thereafter each country in turn could reciprocate with a reduction, with a spiral of less armament each year. Something like this occurred with reduction of nuclear missiles by the United States and Russia in the 1990s.

Global Fuel Control

Starting in the twentieth century, the first nations to couple the new reserves of fossil fuel energy to military technology were able to overrun other nations and cultures with dreadful casualties. Territorial boundaries that were previously adjusted by minor defensive wars turned into wars of conquest as those groups with the fuel consumption technology invaded other nations around the world. In the Second World War, access to fuels was decisive in Europe, and the struggle in the Pacific was partly a struggle for access to oil.

A more peaceful pattern of nations is emerging, with fossil fuels distributed by global markets. Japan achieved economic predominance through development by fuel use and technology that it was unable to do in the Second World War. However, more international equity in access to the remaining fossil fuels may be essential to this new world order.

In the 1991 Persian Gulf war, a new global issue was at stake—whether a country could monopolize the Earth's energy reserves just because those reserves were under the ground of the country's regional landscape. A precedent was set when an international force went to war to protect free access to the great fuel reserves of the Middle East, which were taken over by one country bent on conquest of oil areas. The precedent put access to global nonrenewable energy reserves under international protection. As long as the fuel reserves of the world are generally available at similar prices, no single regime has enough power to start a major war that would require other nations to divert all their resources for survival, as with the world wars of the twentieth century. No one country now has the kind of fuel reserves necessary to start and win a world war.

United Nations Voting

Decisions made by international organizations such as the United Nations might be more successful and settle more conflicts if restructured so that votes corresponded to real empower (measured with annual empower evaluations). Voting can be weighted according to *emergy reserves* and *empower*. Votes that represent real power can prevent the conflicts of miscalculation. Financial assessments and contribution to peace keeping could be based similarly. Something like this was intended when the Security Council of the United Nations was given special powers, recognizing the greater wealth and military power of a few nations at that time.

SUMMARY

As the growth melee strips the remaining natural capital of the Earth, and civilization reaches its zenith, the exclusive dominance of large-scale capitalism can be replaced with an emphasis on cooperation with the environment and among nations. International trade and loans can be made equitable with emergy evaluations. Increased efficiencies are likely to limit tourism, international waste disposal, and extremes in the distribution of real wealth.

A major change in mechanisms for international order is evolving that can replace the old system of territorial defenses. Global sharing of information and increased trade are joining the centers of civilization in common enterprise. Declining resources diminish nations' inclinations and military capacities to encroach on others. Providing that the remaining fuel resources are shared in open markets, great wars of national competition, growth, and conquest may be history. Small conflicts and boundary disputes may be within the power of international organizations to limit.

For a peaceful transition we need to share information internationally rather than sell it, arrange trade and loans with emergy-based equity, and substitute environmental mutualism for resource exploitation.

10

ENERGY SOURCES

For two decades the *energy limits to growth* have been debated fiercely. How much more growth is possible with the energy resources still in the ground? What are the alternatives? Fuel shortages and rising prices followed by surpluses and dropping prices keep people confused about whether energy is important or not. The predictions of rising fuel prices and turndown made after the 1973 OPEC (Organization of Petroleum Exporting Countries) oil embargo were too early because new natural gas reserves were found in many countries. Since then, the main factors affecting fuel prices have been the decisions to export among fuel source countries and the different use rates by the up-and-down world economy.

The economic vitality of a technological society depends on having at least one rich primary source of energy capable of supporting other functions (see Chapter 6). We evaluate each energy source by how much more wealth it contributes to the economy than it uses in the processing. How much can alternate energy sources substitute for coal, oil, and gas in the future? We use emergy in this chapter to evaluate the sources for fuels and electricity, and the alternate energy sources proposed as substitutes.

Contemporary Views on Energy

The neoclassical economics point of view that prevails in advice given to governments predicts that new technology will allow one resource to be substituted for another indefinitely. Some authors cited in Chapter 3 believe in unlimited growth because they believe in alternative sources and/or increasing efficiency in energy use.

Certainly there has been a steady sequence of substitutions of one fuel after another (from wood to coal to oil to natural gas). Each year there have been discoveries of new fuel reserves, but the rate of discovery of really large reserves has decreased markedly. Each year the prospecting and pumping of fuels has to go deeper and finds more dry holes. The net contribution (yield minus effort) decreases each year. After natural gas becomes scarce, the world can use the thicker and then the sparser coal deposits. Along this progression the net emergy from fuels decreases, eventually reaching the point where renewable wood is the best remaining fuel. Even the energy specialists cited in Chapter 3 differ about what can be substituted and how much civilization can be supported.

The ratio of fuel use to gross economic product has decreased since the 1973 oil embargo, and many cite this as evidence of increased energy efficiency and decreased dependence on energy. However, at least part of this change is the increased circulation of money for the same real wealth. As the economy became more complex and urban, money accelerated its circulation in finance and urban processes that are not part of the consumer price index.

EVALUATION CONCEPTS

Much of our civilization is based on heat engines that use fuels to propel cars, airplanes, power plants, and manufacturing, but the fuels are not equal in their ability to drive these heat engines. The higher the temperature that can be developed in the engine, the more useful work that can be obtained.[1]

Temperature of combustion depends on the fuel concentration and air draft. Higher-quality (higher-transformity) fuels are those that can burn at higher temperatures. In order of increasing temperature at which they burn we can list: green (moist) wood, low-density dry wood, high-density dry wood, charcoal, coal, coke, oil, natural gas, and hydrogen gas. But it is wasteful to use a high-temperature fuel when a lower-temperature fuel will suffice. However, we can't tell if a fuel is good enough to support the economy just from its quality. We also have to evaluate its quantity and what is required to mine and process the fuel and restore disrupted environmental processes.

Emergy Yield Ratio

As explained in Chapter 4, emergy measures the contributions of nature and those from the human economy on a common basis. We diagram each fuel system to identify all the inputs, and then evaluate the emergy yield ratio (Figure 6.5). Examples of this indicator are given for typical energy sources in Table 10.1. The larger the ratio of emergy yield to emergy used in the processing, the more other parts of the economy can be supported. If associated environmental losses are included, the yield ratios are less.

Table 10.1—Typical Emergy Yield Ratios of Fuels and Environmental Products

Item	Emergy Yield Ratio*
Dependent Sources, No Net Emergy Yield:	
Farm windmill, 17 mph wind	0.03
Solar water heater	0.18
Solar voltaic cell electricity	0.41
Fuels, Yielding Net Emergy:	
Palm oil	1.06
Energy intensive corn	1.10
Sugarcane alcohol	1.14
Plantation wood	2.1
Lignite at mine	6.8
Natural gas, offshore	6.8
Oil, Mideast purchase	8.4
Natural gas, onshore	10.3
Coal, Wyoming	10.5
Oil, Alaska	11.1
Rainforest wood, 100 years growth	12.0
Sources of Electric Power, Yielding Net Emergy:	
Ocean-thermal power plant	1.5
Wind electro-power, strong steady wind regime	2–?
Coal-fired power plant	2.5
Rainforest wood power plant	3.6
Nuclear electricity	4.5
Hydroelectricity, mountain watershed	10.0
Geothermal electric plant, volcanic area	13.0
Tidal electric, 25 ft tidal range	15.0

*Emergy yield divided by emergy of all inputs purchased from the economy including goods and services but not counting environmental losses (Figure 6.5). Net emergy calculations are from *Environmental Accounting, Emergy and Decision Making* (H. T. Odum 1996).

When fuels were purchased at world market prices before 1973, the ratios of emergy yield to that in the payments for fuel were 40 to 1 or higher.[2] The worldwide energy crisis in 1973 developed after an oil price jump from $.50 per barrel to $34 per barrel arranged by a temporary monopoly of OPEC. Since the 1973 increase of fuel prices, the economies that purchase their fuels have been based on lower emergy yield ratios of about 6 to 1.

Some analysts who tried to evaluate the yield ratios of alternate energy sources used energy instead of emergy. Many of the yield ratios from these evaluations were too high, because they underestimated the large amounts of emergy used in the processing, especially that contributed by human services. Because of wishful thinking and the errors in measuring by energy instead of emergy, millions of dollars were invested in alternative energy sources that had poor net yields. For example, large investments in oil-shale extraction failed, with a multi-billion-dollar loss to oil companies and governments.

Fossil Fuels Versus Electricity

Converting coal or other fuels to electricity requires about four calories to make one calorie of electricity. Most fossil fuel power plants use three calories of coal directly and one calorie of fuel indirectly in plant construction, services, and supply of goods and equipment.

If coal is to be converted to electricity for a city at a distance, converting it first and then transmitting electric power over the distance saves energy. Electricity (at higher transformity) is more flexible than coal and can operate more kinds of machinery. A third of the U.S. fuel use goes toward making electricity.

Transformity and Appropriate Use

A transformity principle given in Chapter 4 was that high-transformity energy should not be used for purposes in which lower-quality energy will suffice. Using electric power instead of fuel for space heating was the example given (because four calories of fuel were required to make one calorie of electricity). However, if electric radiant heaters and electric blankets are used to direct heat only to the human body without general heating of the house, then electric heating pays. In that case the ability of higher-quality energy to be directed and controlled saves energy.

FOSSIL FUELS

Much of the industrial-based civilization of this century runs on fossil fuels. Coal really started the industrial revolution. Early development of coal helped the British Empire predominate in the nineteenth century. Heating gas was made from coal before much natural gas was developed. Then oil became predominant. Although areas without oil have substituted other sources such as coal, oil was the most important fuel supporting most of the twentieth century's developed world. In the 1990s natural gas was being used more and more as pipelines were developed for transport. Coal was being used again in power plants.

For fossil fuels, the richer the deposit, the less work has to be purchased to get and process the energy. On the open market where the suppliers are in competition to produce intense heat for the engines of industry, the richer deposits displace other energy sources because they cost less. In other words, fuel sources with high emergy yield ratios are more economical. In general, the higher the emergy yield ratio from a source, the more heat is available per dollar. For example, biomass (wood, grass, or corn stalks) is not economically competitive with fossil fuels at present because biomass processing requires more emergy and dollars from the economy, even when subsidized by governmental programs.

Going deeper in the ground and further offshore raises costs very rapidly. To go twice as deep nearly quadruples costs because so much extra work has to be done against gravity and friction. Therefore, the poorer and less-accessible deposits of coal and oil have much lower emergy yield ratios. Their stimulus to the economy is less. As deposits in many countries were used up, the world became even more dependent on the remaining few countries with rich deposits near the surface.

Because the deposits with the best net yield tend to be used first, each passing year leaves poorer remaining deposits with increased costs and lower yield ratios. Some estimates suggest that the economic input for getting and processing fuels increases, on the average, one percent per year. This means that the energy available for other purposes decreases by 1 percent each year. The net emergy decreases make fuels less able to support further economic growth.

Coal

Heat and pressure of long-term geological processes convert peat into lignite and coal. Thick sedimentary deposits near the surface have high emergy yields, but most coal is in poorer deposits. With oil and gas becoming scarce in many areas, coal is again being substituted, especially in China and India. Good deposits of coal such as some in the northern Great Plains of the United States may be mined and shipped a thousand miles before the emergy yield ratio drops below the 6-to-1 ratio of typical oil. Much of the available coal has a high content of sulfur, which becomes sulfuric acid on combustion, causing acid rain pollution. The emergy used to scrub out the acids with technological devices accounts for about 10 percent of the fuel used.

Oil

About 16 percent of the emergy of global oil is used to get the oil, about 10 percent is used in transporting it, and another 10 percent in refining it into gasoline, jet aircraft fuel, heating oil, and other products. The emergy of these changes was not included in the examples of Table 10.1. New refineries have been opened in those countries still rich in oil, and refineries have been closed in Texas and other areas that used to be the most important suppliers. Building oil refineries closer to oil sources saves cost on bulk petroleum shipment. Then the higher-transformity co-products are sold to the users at higher prices.

"You're kidding — this is the stuff they wage wars over ?!"

From EARTHTOONS by Stan Eales. Copyright © 1991 by Stan Eales. By permission of Warner Books Inc.

Motor Fuels

Among the first substitutions necessary as oil becomes expensive is something to replace liquid motor fuel for automobiles. Liquid fuels may be made from coal with net emergy yields, but nearly half of the fuel is used up in the transformation process. When imported oil was not available, many of the cars in South Africa ran on gasoline from coal. Already in the United States there are trucks and buses running on natural gas carried in high-pressure tanks. Electric cars and motorbikes are being offered for sale. However, electric cars depend on both cheap sources of electricity and the progress in making light, cheap, long-lasting batteries.

Using electric power for information processing (high transformity) is more important than using it for transportation, as long as there are alternatives. Where the electricity has to be made from fuel, it is cheaper and more efficient just to run the vehicle on the fuel. However, in densely populated cities, electric transportation helps reduce air pollution. For example, electric cars are being tested in Los Angeles.

Natural Gas

Natural gas was ignored for many years, often allowed to burn as it emerged from oil-drilling operations. But now we know that natural gas is more flexible, more easily transmitted in pipelines, and may have a higher transformity than oil. Its use is increasing as more and more deposits are found and pipelines constructed. Because of the high emergy yield ratios of fuel coming in pipelines from Canada, more stimulus was supplied to the U.S. economy than was received by Canada in payment in the year 2000.

Some gas wells also collect liquid fuels formed by gas condensation. Where gas fields are a long distance from the users, natural gas is sometimes transported at high pressure in refrigerated ships as LNG (liquified natural gas). Because of concerns about gas release and explosion, these ships are not permitted in many urban harbors.

Hydrogen

Hydrogen gas is very hot when burned and has a high transformity, but is very scarce in nature. It can be made from natural gas, but for transportation this is a wasteful extra step, as cars and trucks can be run directly on the natural gas. An excess of electricity from hydroelectric or nuclear sources could be used to make hydrogen by electrolyzing water, but this is an added emergy loss since the cars can be run on the electricity directly. Because it has the lightest weight of all molecules and a very hot flame, hydrogen gas is used to propel rockets into space against gravity. It is dangerous because of its explosive combustion with air as demonstrated to the world in the loss of the Hindenburg.

Fuel Cells and Hybrid Cars

Several hybrid cars powered by combinations of fuel and electric power were sold in the year 2000. These cars were offered to conserve fuel and decrease air pollution. However, more emergy and money is required for the more complex equipment.

The deceleration-charging car slows down by letting the energy of the momentum charge batteries instead of dumping that energy into the heat of friction of typical brakes. This car has a fuel engine and an electric motor for driving the wheels. Costs are high, but there may be a net emergy yield.

Fuel cell cars run on hydrogen gas reacting with atmospheric oxygen to generate electric power. Instead of burning the hydrogen, there is an electrochemical interaction across a membrane. In other words, a gas-renewed battery process makes the electricity that operates the motor and drives the wheels. However, the extra step of making fuel into hydrogen and the heavy tanks required to store and carry it must be taken into account.

Another experimental car has a type of fuel cell that can use natural gas directly without first converting it to hydrogen, but it requires a very high temperature (1,000°C or more). In addition to the high emergy required for heavy tanks and complex equipment are the special problems and hazards of high-temperature machinery under the hood. Whether fuel cell cars can generate a good net energy is uncertain.

NUCLEAR POWER

Of several kinds of nuclear reactions that can generate heat and electric power, the nuclear industry developed and standardized hot water *nuclear fission* processes that operate more than 400 plants around the world, 120 of them in the United States. But the dream of unlimited energy has been to develop *nuclear fusion* reactions on Earth like those in the sun.

Electric Power from Fission

A nuclear power plant with nuclear fission develops a temperature of 5,000°C in its core. But much of the heat has to be dispersed to the environment until the temperature is low enough to operate mechanical turbines to make electricity. We live at about 25°C and most of our heat engines—structures of steel, titanium, and aluminum—operate at about 1,000°C. The high temperatures in nuclear reactors would melt some of the materials used in nonnuclear power plants, so special machinery is necessary. Nuclear reactors would be more useful if the large quantity of excess heat released could be used. However, such heat has low transformity and is costly to transport, and few people or industries want to be close to a nuclear plant.

Before uranium is ready to be used in a power plant to produce electricity, large economic resources are used in prospecting, mining, and processing uranium ore into fuel rods. Large quantities of electricity are required to enrich uranium so that the small percent of uranium that is the right isotope is concentrated above critical mass, the level needed for fission. Construction, operation, and decommission of nuclear power plants all require large resources because of radioactivity, the poisonous nature of some of the chemicals involved (such as plutonium and uranyl fluoride), and the potential risks of meltdown if reactions go too fast. Enormous concrete and steel structures, extensive control and safety equipment, highly trained operators, a large governmental supervisory organization, and huge cooling systems are required from the economy.

Nuclear power plants are supposed to last forty years, but questions of safety and the high standards required for operation caused some plants to be closed earlier. After a nuclear power plant is closed, it cannot be taken apart and removed easily. Its inner parts are intensely radioactive with substances that take many years to decay to safe levels. Its giant concrete building is not readily removed or converted to other uses. Some may be sealed off and left to stand with their used, intensely radioactive fuel rods sealed within. The radioactive fuel rods can be removed and processed to recover plutonium for more operations, but the cost is very high and there are associated dangers. (See the following section on Breeder Reactor.)

Money-based evaluations of nuclear power underestimate the emergy contributions to the economy, since there are more emdollars in the electricity generated than in the money used for human inputs. Economic evaluations also underestimate the environmental damages.

For example, each year sees increasing estimates of the premature deaths, disrupted lands, and loss of production due to the Chernobyl nuclear explosion in the Ukraine.

When all the resource requirements that have to come from the economy were considered with emergy in 1975, U.S. nuclear fission was found to generate electricity with an emergy yield ratio of less than 3 to 1. This ratio was little better than fossil fuel power plants for generating electricity (2.5 in Table 10.1). At first the nuclear plants were in operation only 50 percent of the time because there were so many breakdowns, tests, and safety tests to perform. Small leaks in a nonradioactive power plant can be ignored for a while, but not in a nuclear power plant.

Later efficiencies of nuclear power gradually increased as various problems were solved. By 1991 few plants were being built, but the 120 in operation in the United States had become more efficient. When the entire United States nuclear power contribution was evaluated in 1991,[3] a much higher emergy yield ratio (4.5 to 1) was found. The evaluation included storing radioactive wastes permanently on site in the concrete shells, decommissioning, and a share of the emergy lost so far in accidents worldwide.

We conclude that nuclear fission reactors are highly competitive for the part of the U.S. energy budget that needs to be in the form of electricity. Nuclear power has been successful as a major source of electric power for the economies of Sweden and France as well. However, to be competitive as a general source of heat for industry, the emergy ratio has to compete with the fossil fuels' emergy yield ratio of about 6 to 1. As already explained, using high-transformity nuclear electricity to generate heat for engines is also a misuse of electricity.

At the present rate of use, good uranium ores may run out before fossil fuels. Each year, lower-quality ores are used that require more and more of other resources to process. Older plants require more maintenance. Thus the emergy yield ratio of nuclear power plants may decrease again. Nuclear plants have very high capital costs and interest payments. They may have to be smaller if replaced in a time of deflation and fewer resources.

Breeder Reactor

Plutonium is a by-product of the nuclear fission process and, if extracted and concentrated from the spent fuel, can be used to generate more heat in additional nuclear reactions. The by-products from the initial fission reaction are said to "breed" more reactions. The idea is to get more energy from the uranium than with the first fission process alone. However, the intense radioactivity of the spent rods requires expensive robots for the recovery process. Protecting workers from plutonium, which is very toxic and can cause bone cancer, is expensive. Plutonium can be used by terrorists to make atomic bombs with relatively little technology. Breeder reactors were operated in France and Japan but not in the United States. Costs were too high and none were operating in the year 2000. The net emergy, if any, needs to be evaluated.

Nuclear Fusion

Nuclear fusion, the hydrogen bomb reaction, is even hotter than fission, operating at 50 million degrees. It is the same reaction going on in the sun, where enormous gravity keeps the sun from blowing apart. Experience with devices so far shows that without the sun's gravity, vast resources (energy and materials) are required to operate fusion and deal with the high temperatures needed. Judging by the experimental equipment, much more emergy has to be fed back from the economy than was the case in nuclear fission at an equivalent stage of development. This raises doubts that fusion can yield net emergy. So far a billion dollars' worth of apparatus have been required to generate a few seconds of fusion.

RENEWABLE SOURCES

We have been discussing the emergy yield ratios of nonrenewable resources—those that our civilization is using up faster than the geological processes are making them (see Chapter 4). Since the reserves of nonrenewable sources supporting the economy are gradually running out, there are many proposals to use renewable energies as alternatives. Renewable resources are those that are steadily supplied and continue to be available even as they are used. For example, the sunlight falling on the Earth each day comes regularly whether it is used in agriculture, forestry, or photovoltaic cells. However, it is a dilute energy spread over a large area (low transformity). Some renewable sources, such as hydroelectricity generated from the water that falls in mountains, have high emergy yield ratios (Table 10.1). However, these are not extensive enough to support the global economy that has developed on fossil fuels.

Diverting Natural Sources

It is usually assumed that a newly developed energy source is a new contribution to an economy. However, energy flows found in nature already carry out works that contribute environmental values to society whether humans control them or not. A consequence of the maximum empower principle (see Chapter 4) is that nature's self-organization process uses available energy to the maximum possible extent. Because economic analysis cannot evaluate the real wealth contributions of the environment, developments frequently divert high-quality energy from natural processes without regard to the role these energies have in the landscape as a whole (contributing indirectly to the economy). Any calculation of the utility of a new or proposed energy source should include the former uses lost when the source is put into new use. This leads to a general policy:

> Before diverting environmental resources into new economic uses, existing emdollar contributions of environmental processes should be compared with the dollars to be obtained by the development.

Elevated Water

The heated ocean and atmosphere collect and transform the energy of sunlight as a giant low-temperature heat engine that causes winds, currents, rains, and snows over land. The energy of atmospheric weather systems is more concentrated than that of the sunlight that made it, but most of the sunlight was used up in the process of conversion. The rains and snows falling in mountains are a further concentration of energy into the potential energy of elevated water and glaciers. The rivers that result operate the geological process of erosion and deposition, parts of a cycle that contours watersheds and rebuilds mountains (Figure 9.1). By constructing dams, the energy of the elevated water can be converted into hydroelectric power. The quantity of energy converted is small compared to the initial sunlight but high in transformity.

Rivers dammed for hydroelectric power are diverted from their geologic and biologic work such as the support of migratory fisheries. Torrential streams in southern New Zealand, by crashing rocks together, used to make soils that enriched the valley agriculture. After that energy was diverted into electric power to make aluminum ingots for export, the soil fertility has had to be maintained with costly fertilizers.

Salmon are a high-quality (high-transformity) food source that draws emergy from watersheds and from the sea in the course of their life cycle of migration and spawning. The demand for hydroelectric power and information development is in conflict with the movement to eliminate dams that block salmon runs. (See Chapter 15.)

Geothermal Energy

Any difference in temperature can be used to do some work. The larger the difference in temperature, the more work that can be obtained. Temperature increases gradually with depth in the earth. Except in volcanic regions, the difference in temperature is too small for human needs.

In Iceland, a cold area with active volcanoes, hot geothermal waters are piped to heat the city of Reykjavik. Geothermal heat in New

Zealand operates hot springs used by the Maori for heating and cooking and for tourism. In this type of use the geothermal energy is steady (renewable). In other areas, pipes were put down and the hot waters used for geothermal electric power. After a time the ground was cooled and the previous uses were lost. As the heat declined, pipes had to be moved to other hot areas. In other words, geothermal heat is not renewable when used intensively. In 2000, Iceland announced a plan to use its volcanic thermal power and mountain water power to make hydrogen for transportation and other needs. This plan may waste energy by making a higher-quality energy than needed.

The surface waters of the tropical seas are warmer than the deeper waters by up to 25°C. This difference can be used to operate a heat engine, but some kind of ship with long pipes is required. The attempt to use this temperature difference is called OTEC (ocean thermal energy conversion). The source is not strong with so small a temperature difference over such a long distance. Corrosion and tropical storms make such systems short-lived and costly to maintain and replace. If the installation is on land at the seashore, costly long pipes are required to reach the deep, cooler waters. This may not be competitive, because the emergy yield ratio was estimated to be less than that of wood.

Tidal Energy

There is stored energy in the orbiting of the Earth, sun, and moon left over from the time the solar system formed. Some of that gravitational potential energy pulls at the oceans to make a wave of tide that bounces back and forth like water sloshing about in a bowl on a train. The tide makes estuarine currents, forms beaches and channels, supports life cycles of fisheries, generates salt marshes, disperses wastes, and generates shellfish beds.

In a few places, tidal energy has been diverted to generate electricity by damming an estuary and letting the water in on rising tide but making it drive turbines on the way out again. Where the tide is a very large one, there is a good yield ratio. One successful tidal power station is in La Rance, France, where the tides average twenty-five feet (Table 10.1). However, the net benefit is not so favorable when the loss of the estuarine fisheries and the former natural waste purification service is included in evaluations.

Water Wave Energy

The trains of waves breaking on the seashore drive shore currents that transport sand, aid life cycles of marine life, maintain the beaches, and process coastal waters through the sands that filter the water. High-energy beaches are cleansed and renewed daily. They are a focus of human life and recreation. Major economic values depend on the free work of waves reaching the shore.

Some devices, such as buoys that generate their own electricity for their lights, take energy from waves, but harnessing waves is difficult because of the great range of energy during storms. Displacing wave energy away from existing coastal work for power purposes may not yield net emergy benefit.

Wind Energy

Without requiring money, wind energy already contributes to the economy by operating the earth's atmospheric-oceanic system. Winds stir the seas, are part of the rain delivery system, help energize the transpiration and gaseous diffusion that makes vegetation and crops grow, and operate geological weathering that builds soils. Winds are used more directly in the economy to dry laundry, hay, and food products; blow waste gases from smoke stacks; and help cooling towers disperse waste heat.

Although wind energy is more concentrated than sunlight, it is spread out enough to require costly structures with large surfaces to catch the force of the air and generate mechanical energy. Available wind energy increases with the cube of the velocity (velocity multiplied by itself twice). Low-velocity winds have little energy. High winds have tremendous energy, but when winds are too strong and turbulent, sails or windmill blades have to be disconnected to prevent damage to equipment. In many areas of the world the winds of passing storms alternate between being too light to yield net emergy and too strong to operate wind technology safely.

Where winds are favorably steady and moderately strong in some mountain passes in California, hundreds of large, electricity-generating wind turbines are arranged in rows called "wind farms." These modern wind rotors have taller and larger rotating blades on towers to reach stronger winds further from the ground. Management by computers has made it easier to adjust the load on the windmill blades to the variation in the winds. Their costs in emergy and money are large, but they do yield net emergy (Table 10.1).

Before the fossil fuel era, wooden sailboats and Dutch-style windmills prevailed in coastal areas where winds are 20 to 40 percent stronger than over landscapes further inland. The wind energy was used for sailboat transportation and mechanical work such as grinding grain. More emergy was yielded than was required in human services. More of the emergy of the yield came from the great timbers used in windmills and in the sailing ships to catch the wind than from the wind energy itself.

SOLAR ENERGY

Solar energy on the scale of human society is dilute, but very practical where it can be used for low-quality purposes without much cost,

HAGAR THE HORRIBLE by Chris Browne. Reprinted with permission of King Features Syndicate.

such as gentle heating. If walls are already paid for and built for some other purpose, they can be arranged to catch sunlight for house heat, and there is a net benefit. Sunlight can be used along with wind for efficient drying. These are sometimes called "passive" solar energy uses.

Much of the solar energy falling on the Earth is caught and converted to biomass by the green plants whose cells operate photosynthesis. The plants build and maintain their own systems, with thousands of varieties covering the Earth with green converters. In the deserts where the green cover is sparse, the solar energy generates winds and does geological work on the soils and landscape. Much research has gone into trying to use the solar energy more directly. One approach uses photosynthetic products.

Biomass and Growth Time

Because of the dilute nature of sunshine (spread-out rays), the net emergy of all solar energy conversions to higher-quality energy is small. The biosphere concentrates the emergy of sunlight by making organic matter with photosynthesis and then accumulates the organic matter, transforming it further in food chains. Green plants using many biochemical reactions convert solar energy to biomass with a small net emergy gain. Green plant cells after a billion years of evolutionary trial and selection may have achieved the optimum efficiency that maximizes output (see Chapter 4).

The key factor is the accumulation time of the solar-conversion products. The longer the accumulation, the higher the net emergy yield.[4] Wood varies widely in its energy content as a fuel. Young, rapidly grown wood from some plantations has low density, and is neither concentrated enough to maintain a very hot fire nor hot enough for a power plant. However, older, denser wood with more chemical substance, when dried, forms a good fuel.

Forests that grow on their own take little effort from the economy. Given a hundred years' growth, emergy yield ratios of wood may be 4 to 1. Their growth is slow because most of the energy goes into soil production, diversity, and forest maintenance.[5]

When wood biomass is grown in forest plantations for shorter times–ten or twenty years–and the products collected in one place, the emergy yield ratio is about 2 to 1, less than forests given more time. Most biomass grown as agriculture has even lower emergy yield ratios. Biomass grown for only a few years can not compete as a fuel so long as the emergy yield ratios of fossil fuels are 6 to 1 or higher.[6]

Crop biomass such as sugarcane and corn has been transformed into ethanol and used as a motor fuel. Methanol can be made from wood. However, the emergy yield ratios are barely greater than 1 to 1. This means that agriculture can supply fuels for cars, but it will not compete with most other fuels. It will not stimulate the economy to operate transportation with these fuels.

The net emergy yield of biomass from natural forests and agriculture was enough to support human settlements in past centuries before fossil fuels were important. Consistent with the concepts of pulsing in Chapter 5, there were rises and declines in populations and cultures. When areas were lightly populated, biomass in soils, trees, and wildlife could accumulate, providing the means for a later surge of consumption and a brief flowering of civilization.

Peat
Dead organic matter that accumulates in swamps and bogs is called *peat*. It represents long periods of storage accumulation and thus can provide net emergy to users. It is widely distributed in wetlands the world over. Wet peat is not a good fuel in ordinary furnaces because it takes so much of the heat produced just to evaporate the water. Peat is a good fuel when sun- or wind-dried or when used with special furnaces that recondense the water vapor to recapture the heat used in evaporation.

In Ireland, peat is used for power plants after it is dried in summer sunlight. However, peat deposits may be serving the economy better in other ways such as filtering and purifying natural water. Mining it for fuel may take away as much emergy value from the economy as it contributes as fuel. Examples are the peaty wetlands that control water quantity and quality in many Florida rivers.[7]

Solar Technology
Many futurists hope that our new postindustrial society is going to be able to run on solar technology (see Chapter 3). Although there

are enormous outflows of energy in the sun, the energy spreads out when it flows to Earth and has to be reconcentrated to be used for high-quality purposes. Wherever accumulated, the sun's rays do many valuable services, but most of the uses do not yield net emergy. The solar technological hardware devices directly and indirectly use more emergy in resources and services in their manufacture, operation, maintenance, cleaning, and orienting to sunlight than they produce in electricity. Just to make a magnifying glass to concentrate the sun's rays takes more work of high-temperature heat than the glass can concentrate in its lifetime. Advocates who quote net e*n*ergy (instead of net e*m*ergy) calculations to claim the opposite use energy without multiplying by the transformities necessary to properly evaluate materials and human services on a common basis. The technological devices that convert solar energy directly to mechanical or electrical energy analyzed so far have no net emergy yield.[8]

When green land plants catch, transform, and concentrate solar energy, the working unit of chlorophyll in plants converts solar photons into electrical charges. The chlorophyll units in green plants are very efficient biological photovoltaic cells. Their products do have net emergy yield. But like the leaves, these electrical charges are dilute—spread out over the landscape. The energy in the charges operates biochemical reactions to make sugars, which are still dilute. Then plants concentrate the sugars by moving them from the leaves to limbs, trunks, and roots. There the plants concentrate the organic matter by converting it into wood fibers, concentrating biomass further by adding a little each day. The point here is that solar energy must be concentrated as well as converted to be used, and much of the energy first converted is used up in the process.

The conversion of photons to electric charge is also done with human-manufactured hardware—photovoltaic cells made of silicon, selenium, and other substances. But the charges on the hardware cells are collected on wires as low-voltage electricity (.5 volts). The electricity must be further concentrated by a network of wires and a series of cells to get 120-volt power. Green plants yield net emergy in making biomass, but the hardware cells making electricity do not. The goods and human services to produce the solar cells use more emergy than is in the electrical yield in the life of the cell (Table 10.1).

The hierarchical nature of energy may explain why. It appears that energy is more efficiently transformed from low quality to high quality when it is done in a series of steps (see Chapter 4), instead of trying to skip steps. Low-transformity sunlight is more efficiently transformed to high-transformity electricity by converting through intermediate steps of the food chain (sun to plants to biomass to electricity) than by using photovoltaic cells to generate electricity from sunlight in one step. The natural arrangement that uses a series of conversion steps is apparently a better net emergy yield than a single-step process.

The electric charges from green plant cells can also be collected to form electric current. Years ago we used algal mats, which have a half-volt charge from top to bottom, to make solar batteries.[9] However, there was little prospect of a high net emergy because of the costs of the wires necessary to collect and concentrate the charges from the wide area required to receive the sunlight.

The real limitation may be a fundamental physical one of concentrating light. The plant kingdom has had a billion years of evolution to maximize efficiency of solar conversion and collect the energy in further steps to support higher-quality systems. The green plants may already be at the thermodynamic efficiency limit of conversion of sunlight to higher-quality energy. Efforts to industrialize the sunlight may be near their limit.

Solar-cell research has been well funded for most of the century, and as research has progressed, the percentage of sunlight converted to electrical energy in the hardware cells has increased. Decreases in costs of solar-voltaic cells in recent years encouraged many to extrapolate to a day when there would be a net emergy competitive with other alternative sources of electricity. The efficiency may eventually approach that in the plant cells. The many people pressing Congress for more research–"the solar lobby"–are well-meaning but sometimes misled by the way efficiencies are exaggerated.[10] The correct comparison is between the more-efficient chloroplast and the less-efficient hardware cell.

For sixty years, advocates have promised that photovoltaic cells would be economical when prices of fuel rise. This is a fallacy because higher fuel prices also raise the costs of the necessary services. The net emergy is not improved.

The net yield is less after green plants have used the organic matter from the chloroplast photosynthesis to maintain and operate the whole forest structure required to support and replace the leaves. The net yield of photovoltaic cells is also much less after the energy is used to maintain and operate the structures for supporting the cells, changing their orientation with the sun, and collecting the electricity.

Solar-voltaic cells can be important without yielding net emergy. They are appropriate wherever electricity is needed for portable devices like calculators or for remote situations such as weather stations in a wilderness or satellites in space.

Solar energy is still the basis for our future as it was for our past, even though solar technology will be only for special purposes. Just imagine how long we would exist if the sun turned off. For transition and descent we must learn to use the sun more efficiently in the processes that do convert sunlight with good net emergy, such as forestry, fisheries, and agriculture.

The conclusions, based on calculations of the energy required to collect solar energy in various ways, is that biomass is the only proven way to run our society on solar energy. However, on a renewable basis, solar energy will support only a fraction of the present levels of population and industry in developed countries.

ENERGY CONSERVATION

Measures to save energy in the use of appliances–for example, better insulation for houses, more-efficient refrigerators, and heat pumps– often yield net emergy even though the energy-saving equipment adds cost. Some efficiency measures are not economically competitive because they cost more. However, the net emergy of fuels saved is so great, it may make a net benefit to society to provide a financial subsidy such as tax relief for fuel conservation. More net emergy evaluations of energy conservation measures need to be made.

General air-conditioning of buildings made tropical and subtropical areas popular and productive. The electric power used in states like Florida and Arizona exceeds the use by states in cold climates. Openable windows were often eliminated, and the air-conditioning operated unnecessarily. Air-conditioning can be much reduced and buildings redesigned to use more natural cooling.

Pulsed Use

It is economical to run electric power plants continuously, but the demand for electricity goes up with daytime uses and down late at night. Activities in a region can be coordinated to spread uses evenly during the day and night. For example, some hot water heaters are programmed to work only at night. Another idea for saving night power is to pump water into a mountain reservoir, letting it operate hydroelectric turbines when coming down later. However, when "pumped storage" is loaded for maximum power, it has to discharge half or more of the original energy into heat. (See the time's speed regulator principle, Chapter 4.)

Cogeneration

Major energy conservation is possible by combining two or more uses of heat in the same process. The first uses the highest-quality (high-temperature) part of the process and the second uses the lower-temperature heat. For example, many universities have used the waste heat or low-pressure steam from their own electricity-generating plants to heat buildings. Refineries and petrochemical companies that need heat to separate fractions by distillation have added electric power plants to supply their own electricity or sell it to the main local power authority.

Conservation with Solar Heaters

Although they do not yield net emergy, solar technology devices can be useful for energy conservation. For example, solar hot water heaters save the fossil fuel energy that would otherwise have been used. What prevents solar heaters from being widely used is the high initial dollar cost of the equipment compared to fuel costs saved each year. Affluent people are more likely to afford energy-saving devices. Whether a device is a benefit requires a comparison of the emdollars of the fuel saved with the interest the installation money could have earned in the bank.

Solar Energy in a Lower-Energy World

In times of lower energy, solar green production for society becomes proportionately more important. Agriculture becomes less intensive with lower investment ratios (ratio of purchased, fossil fuel–based inputs to free environmental inputs, Figure 6.6). Although proposed for low-energy areas, solar technology with its high emergy investment ratio is not competitive with solar biomass processes. Purchasing the necessary materials and services costs too much.

Solar energy through agriculture, forestry, and aquaculture can supply each of the necessities: food, fiber for clothing, materials for housing, heating, fuels for transportation, chemical stocks, etc. But there is not enough sunlight to do them all at current levels without the fossil fuels. For the times ahead when fuel use is decreased, policies may be needed to allocate parts of the solar landscape to each of the functions as necessary to keep the economy in balance.

SUMMARY

Although many energy substitutions and conservation measures are possible, none in sight now have the quantity and quality to substitute for the rich fossil fuels to support the high levels of structure and process of our current civilization. Whereas ecosystems, forests, agricultural systems, and fisheries using sunlight have net emergy yields to support society, solar technology does not because the dilute nature of sunlight prevents efficient conversion directly to mechanical or electrical energy.

The trend of substitution of one fuel for another continues toward more use of natural gas, but proven fuel reserves are not increasing. Because 71 percent of the whole Earth empower comes from fossil fuels (Figure 6.2), global consumption eventually has to be reduced to less than one-third of its current level. The developed nations that depend on nonrenewable resources for 80 to 90 percent of their energy will eventually have to reduce either their populations or their living standard (emergy use) by 80 to 90 percent. However, with reduced populations we can look forward to a new but smaller agrarian economy, green again, enriched with knowledge developed in the fuel-rich century of complexity.

11

SUSTAINING A NATION

To achieve and sustain a maturing civilization may already be the unwritten policy of nations. Formerly accustomed to growth, people sense a change of direction with concern for the future (see Chapter 3). Principles from Part II help us understand the transition. Wearing a national hat in this chapter, we seek policies for the time of climax while preparing for descent.

According to the pulsing models in Chapter 5, the period at climax is relatively short, a time of transition from growth to descent when it is possible to balance emergy inputs and losses. To extend the summit we need to sustain the national emergy budget and waste it less. We can increase real wealth if incentives are applied to encourage favorable balances of international exchange and energy conservation (see Chapter 10).

For several centuries in the United States, agrarian colonial development by European-descended Americans used up soils and forests and moved west. Then energy consumption shifted to fossil fuels. Resources were wastefully converted, and cities developed at a frantic pace, outdistancing many other economies. During growth, priorities expanded resource uses with the low efficiency and waste that accompany speed. Foreign resources were used more and more as the high-quality reserves in the United States dwindled. Water and lands used with fuels began to limit growth. In the 1990s the gross national product was still increasing by 2 percent or more each year, but some of this increase was *unearned income* circulating in high finance and industrial concentration without producing real wealth. When growth decreases, traditional economic principles, predictions, and policies do not work well, and people may be ready for new policies. An ecological analogy helps us see civilization in overview.

INSIGHTS FROM CLIMAX ECOSYSTEMS[1]

The mature human economy is like the climax state (see pp. 84–85) of a mature ecosystem such as an old-growth forest. As it matures, a natural forest builds live and dead woody structure, increases biodiversity, carries on maintenance and replacement, and recycles materials for reuse. When the quantities and diversity in an ecosystem are as great as can be supported by the resources available, net growth has to stop. For a time the mature system's main product is maintenance. It replaces parts without any net increase. Later the ecosystem starts over after its structure is removed either by mechanisms within the ecosystem such as tree falls, insect epidemics, and fire, or by pulses from the larger system outside such as hurricanes, landslides, and clear-cutting.

Mature ecosystems are more effective at using all the available resources than early stages of growth. More working consumers are specialized, increasing the division of labor. For example, each plant species is chemically different, and all are eaten and controlled by specialized insects with adapted digestive processes. At climax the prevailing species cooperate more in use of the resources. Insects occupy different "niches" (occupational roles) so that they don't compete much. The system is more complex. Genetic information increases in the progression to climax, but there appears to be an information limit. Comparison of mature ecosystems shows similar quantities of DNA (genetic matter) per area[2] and similar species diversity (analogous to diversity of occupations).

Human economies also tend to increase in quantity and complexity until they are using available resources as fast as they can be supplied. The human economy also develops mechanisms for maintenance, recycling for reuse, and diversity. Cities have many kinds of occupations, each with a different specialty, making a cooperative and efficient division of labor.

THE CONDITION OF NATIONS

With the rest of the world devastated by World War II, and with abundant resources available at home and from import, the U.S. economy dominated from 1950 to 1970. Most U.S. citizens enjoyed rising wages and affluence in cars, homes, and creative activities. Its economy then was stimulated by oil with an emergy yield ratio of 40 to 1.

After 1973, the net emergy of fuels on the world market decreased, ranging between 3 to 1 and 12 to 1. Strong competing economies, based on imported fossil fuels, developed in Europe. Others developed around the rim of the Pacific. Growth continued but not as quickly.

In the 1990s, lower-priced manufactured products from abroad forced many American manufacturers to lower wages by not giving raises, by contracting work out to businesses with lower wages, or by moving their work overseas. More and more products used by Americans came from abroad through giant distribution corporations like Wal-Mart, Kmart, and Office Depot. Competition from foreign food imports threatened U.S. agriculture, but kept food prices low. American consumers got marvelous bargains. The imported goods were so cheap that the standard of living (emergy per person) in developed countries was sustained without much inflation. However, the distribution of wealth was uneven (see Chapter 12). Many underdeveloped countries were devastated by the removal of their resources with payment received only for the human labor used in the processing (see Chapter 9).

The United States in 1997

In Figure 11.1, the United States in 1997 is shown in overview with the circulation of money above and real wealth inputs below. In 1997, the gross national product was 8.1 trillion dollars per year (Figure 11.1a). The inflow of foreign dollars was about 19 percent of the gross national product. Below in Figure 11.1b is the flow of real wealth evaluated with emergy. The basis for the national economy was 22.4×10^{23} solar emcalories per year, of which 28 percent was supplied by imports. The United States received 2.5 times more wealth in imports than it exported.

Figure 11.1. Diagrams of the economic-environmental system of the United States in 1997. (a) Circulation of money, with data from the U.S. Statistical Abstract for 1998; (b) annual emergy flows calculated from 1997 energy data with procedures in H. T. Odum (1996, Appendix D).

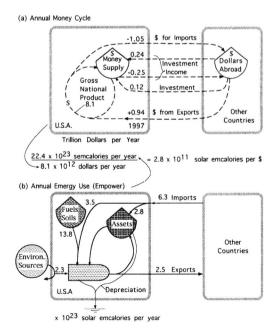

NATIONAL ENERGY POLICY

Getting the most out of an economy means getting the most net emergy from resources, both imported and domestic, and using them efficiently. People think fuels are of minor importance because fuel dollar costs are only 10 or 15 percent of the gross economic product. But because of their high emergy yield ratio (see Chapter 6), fuels are 80 percent or more of the national emergy budget. They typically contribute six times as much real wealth as they cost.

Earlier in this century when other countries were less developed, the emergy yield ratio of marketed fuels was 40 to 1, stimulating rich emergy growth. The U.S. economy represented half of the world fuel consumption and was the strongest in the world. Now, with a lower emergy yield ratio of fuels purchased internationally, the amplifier effect is less. Other countries import fuels with the same ratio, a main factor giving Japan, Hong Kong, Taiwan, Korea, Germany, Sweden, and all the other developed countries vital economies. In 1997, the U.S. share of world fuel use was down to 19 percent of the total. Its power was no longer dominant.

First priority for a nation's policy is to maintain access to those primary fuel sources with the highest emergy yield ratio. National fuel policy should seek sources with a ratio of 6 to 1 or higher, the ratio available on international markets in recent years. As a cheap amplifier, fossil fuels should be made available to all productive enterprises at cost. As long as the availability of primary sources with similar net emergy continues, a mature economy can continue. The Persian Gulf war was in response to a threat to fuel sources of the Near East.

Importing Fuels

Because a country receives more economic stimulation from fossil fuels than it pays out at current prices, buying foreign fuel is a good deal and would be so even if the relative price were double the present one. Policy should not discourage fuel imports. Countries that use their own oil first soon run out of cheap oil so that their economy becomes even more dependent on foreign fuel. Because they use their most-concentrated fuel reserves first, the reserves left are progressively more expensive to mine and process. Their economies diminish as more and more of their expenditures have to go right back into getting energy. With higher prices their products don't compete well.

The so-called energy independence policy of U.S. administrations in the 1970s and 1980s was short-range myopia. Encouraging temporary self-sufficiency used up home fuels prematurely and accelerated longer-term foreign dependency. By using foreign fuels first, countries retain their own reserves for use in emergencies or to keep prices down. In the long run, the total energy available for that nation's use is greater.

The *use foreign oil first* policy allows a country to use foreign oil *and* home reserves.

Nations with oil reserves still near the surface and emergy yield ratios of 50 to 1 or more should use their oil at home to benefit most. Oil-rich Near East countries are developing the refineries and manufacturing to use their own oil more. This may eventually reduce the availability of cheap oil to other countries. Currently they export at market prices that divide the emergy benefit with their customers. As the Persian Gulf war showed, developed nations with military power probably would not allow them to do otherwise.

Cuba is an example of a nation without the money to import fuel in the 1990s after the fossil fuels stopped coming from the Soviet Union. Power plants ran only a few hours a day, and few people could drive cars; the standard of living was reduced.

Fuel Tax

Because of the 6 to 1 emergy amplifier effects, taxes on *productive uses* of fuels and electrical energy dampen an economy much more than other taxes. Since energy runs everything directly and indirectly, anything that inhibits the energy available inhibits the economy. Some underdeveloped countries unwisely levy fuel taxes on fuels used in essential work.

A principal waste in our society is using fuels in nonproductive activity. We drive more cars than necessary, drive them too often, and drive cars with too much horsepower. We use cars for commuting because cities are not organized with alternative transportation. Because higher costs of energy do cause people to eliminate some stupid waste, higher fuel taxes may be needed in the United States for these wasteful uses, with exceptions for efficient use of fuels in agriculture, industry, and essential family transport. In practice it is hard to write laws that discriminate between productive and wasteful uses. In Europe, where taxes on automobile fuels are high, cars are smaller and use less fuel.

Regulation of Automobiles

The love of automobiles with extra power and luxury features has been central to the American culture. Cars save individuals' time, a conservation of emergy if the time saved goes into productive work. However, cars for individuals cause cities to spread out, commerce to organize in strip development, and people to live away from their jobs. Not only is there waste in the excess use of fuel and machinery, there is waste in the organization of the landscape and the commuting of drivers.

A way to eliminate waste is to require permits for cars with excess size and power and for other unproductive resource uses. However,

DAVE by David Miller. © Tribune Media Services, Inc. All rights reserved. Reprinted with permission.

eliminating waste with regulations requires a bureaucracy, which may have its own wastes, inefficiencies, and corruption. Except in wartime, when gas was rationed and permits were required for larger users, democratic nations have avoided regulating resource use.

Policies ought to reduce the use of cars now. Unfortunately, the reverse happened in the late 1990s. When fuel consumption decreased in the depressed economies of Asia, fuel prices dropped. In the United States, families bought large vans and trucks in a very wasteful fad. Although larger size was supposed to provide some protection in case of accident, the greater weight makes it harder to stop and the top-heavy shape causes rollover accidents. All this waste can be expected to change as the nation becomes more efficient. The present pattern can not be sustained when fuel prices rise.

Domestic Resource Use

As explained in Chapter 6 (Table 6.2), raw materials and agricultural products contain several times more real wealth as emergy than is in the money paid for them. The money goes only to the humans for their part of the work. In 1996 the United States, including Alaska, sent abroad large shipments of real wealth as corn, wheat, rice, soybeans, wood, fish, phosphate, and coal, and received less emergy in monetary exchange. However, the net loss was more than balanced by the high net emergy of the fuels used. Energy uses in the United States in 1999 diversified, drawing more equally from domestic coal, natural gas, nuclear power, and imported foreign oil.

Good commodity policy keeps raw products for use within a nation. Home use policy lowers prices of food, housing, paper, and fuel within the country and raises the standard of living of consumers. Less emergy goes for transportation. In Chapter 6 the principle of reinforcing production was explained. The policy for agriculture was successfully implemented for many years by using subsidies. In place of excess production of one crop for export, farmers can diversify their products for domestic use, a pattern that will fit the needs of lower energy times ahead (see Chapter 16). The emergy saved by keeping farm products for the domestic economy justifies additional government subsidy to balance their income and cost.

FOREIGN EXCHANGE

National policies are needed to control the exchanges of money and emergy in exports, imports, and finance that affect national welfare.

Monetary Exchange

In recent years the United States has spent more dollars on imports than it has earned in international sales (Figure 11.1). The dollars that accumulated abroad recycled into the U.S. economy as foreign agencies purchased stocks, bonds, and income-earning businesses and property within the United States. In other words, trade imbalance generated investment from abroad, increasing the U.S. international debt. Thereafter the interest, dividends, and profits from these investments went abroad. Although this was small in one year, over several years this trade deficit caused accumulation of debt. Part of the 1997 federal debt of five trillion dollars was held abroad. By going abroad, interest on this debt carried real wealth from the United States that could have gone into U.S. use. Achieving a monetary balance of trade and reducing debt are policies for sustaining a mature nation.

Each time there is a surge in investments from abroad, there is a pulse of prosperity due to the extra money circulating and increased imports. But later, a dampening occurs when interest paid out reduces dollar circulation and carries emergy out. Good national policy limits investments from abroad.

Even now, money going out for fuel imports is a major part of the balance of trade, which will only get worse as suppliers have to drill deeper. The immediate solution is reduction in fuel uses that don't contribute to productive empower. In order to suppress borrowing and keep trade balanced, a policy of restricting foreign purchases for goods other than fuels may be required eventually.

Balance of Emergy in International Trade

As explained in Chapter 9, developed nations presently enjoy greater emergy in imports (especially those in fuels) than in exports. These nations can sustain their national pattern and standards of living so long as their favorable emergy balance of trade continues. Although the United States has diversified its fuel sources to include more domestic coal and natural gas, it still requires foreign oil shipments with a large net emergy exchange to sustain its present standard of living. Most of the high-emergy-yield oil reserves in the United States are gone. When a temporary monopoly by oil-supplying nations jumped oil prices in 1973, there was a temporary decrease in the emergy yield to developed nations that caused a 15 percent inflation (less emergy was obtained by the United States and more emergy went out in the money paid).

Effects of Free Trade

Prevailing neoclassical economics holds that efficiency is achieved by allowing free markets to eliminate competing businesses that have higher prices. Advances in the technology of communication and finance allowed this economic ideal of competition to be accelerated by agreements for international free trade (NAFTA, North American Free Trade Agreement, and GATT, General Agreement on Tariffs and Trade). The large multinational companies benefited by shifting work to less-developed countries abroad.

To compare the real-wealth buying power of U.S. hourly wages with those abroad, the wages in dollars should first be converted to emdollars using emergy-money ratios (Table 6.1). For example, the real wealth (emergy) purchased with a dollar in China was six times larger than that bought by a dollar in the United States. On an emergy basis, a wage in China of one dollar was equivalent in buying power to a U.S. wage of $6.20. Even allowing for this correction, the pay rate in China was less than that paid middle-class labor in the United States. The increased profits of multinational corporations using cheaper labor stimulated the stock market.

Tariff Alternative

Some reacted to U.S. jobs lost overseas by advocating a cancellation of international free trade agreements and substituting tariffs so as to maintain U.S. wages at the former level. The idea was to become a more isolated, self-sufficient economy. Earlier in the century, tariffs were enforced with federal laws as the United States tried isolating its economy. At that time it had abundant sources of oil emergy within its borders and was not so dependent on foreign imports as it is now. Even then the economy did not prosper and many tariffs were abandoned.

Trade barriers can cause retaliation by other countries that also add tariffs. With tariffs, prices of many products become too high to sell abroad. High prices of necessary imports means a lower standard of living. Less income from export sales is available to pay for the necessary fuel imports on which production depends. With tariffs, monetary balance of trade can be negative, debt can accumulate, and inflation can increase if net emergy is lower because of reduced fuel imports. Inflation can drop the standard of living, and high fuel costs inhibit production.

NATIONAL FOREIGN POLICIES

The following are foreign policies for national benefit.

Treaties for Emergy Equity

There is a better alternative to free trade or tariffs. As explained in Chapter 5, inequitable foreign exchanges make bad international relationships, do not maximize world empower, and are unstable. Instead, international trade treaties could arrange for emergy equity not attainable by free trade. For example, exchanges between the United States and Mexico in the 1990s profited the United States more on a real wealth basis because the fuels and agricultural products it received had a larger net emergy than the manufactured goods sold to Mexico. New exchange agreements between the United States and Mexico and between the United States and Canada could use emdollars to arrange equity.

How can a developed country maintain its current favorable emergy balance and still be equitable in trade with less-developed countries? One answer is to retain a favorable emergy balance through fuel imports, but adjust all other exchanges to balance. This policy assumes that all countries can have access to world fuel markets (the proposed *doctrine of global access to fuels* from Chapter 9). Fuel-exporting nations export more emergy in their fuels, but they receive the profits from sales.

By controlling exchanges of money and commodities across its borders, a less-developed country can restrict outside investments to those arrangements with emergy equity. By limiting unessential imports, a nation can direct its foreign exchange into acquisition of fuels and other inputs to balance its emergy exchange (see Chapter 6).

Regulating Global Capitalism

Controlling global capitalism's inherent tendency for short-term exploitation of resources can prevent it from undermining a nation's resource basis and causing collapse. The discount idea is often used to justify stripping natural capital. Selling a resource now is supposed to provide more money than waiting until later, because the money earns interest as soon as it is in the bank. The fallacy is that the resource left in operation without being removed produces new value as fast or faster. Even in the short run it only pays an economy to exploit a resource if the dollars of interest paid by the bank are higher than the emdollars of new value generated by the resource system. For example, growth by an intact forest may add more new value per year than cutting the forest, selling the wood, and putting the money in the bank for interest. As global growth ends, interest rates are expected to decrease, reducing the economic benefits from exploitation.

At the end of the nineteenth century the U.S. Supreme Court ruled that a corporation had the same democratic freedom as an individual in its delivery of economic impact and investments. This was compatible with a stage of fast growth because it aided the faster-growing competitors. Now, however, in order to restore economic control and power to duly elected national, state, and local governments during transition and descent, this decision will need to be reversed by constitutional reinterpretation or amendment. To maximize performance in the new stages, cooperation and control of destructive competition are appropriate.

Immigration

Without growth, fewer jobs will be available for immigrants than in the past. Restricting immigration helps maintain the standard of living. Fewer people mean a larger emergy share per person. Less money is sent to relatives in the countries of origin. Public perception in California and Florida sees immigrants competing for jobs and public welfare. These states seek federal funds to cover their expenditures on immigrants like the refugees from Cuba. However, highly educated immigrants bring in more emergy than they use as new residents. Therefore national policy encourages only the exceptional immigrants.

Defense

Although the sharing of culture and attitudes through television and the Internet (see Chapters 9 and 17) may be reducing the need for territorial defenses, national safety and territorial boundaries require some defense force. Too much defense weakens the nondefense part of the national economy so that it becomes vulnerable to foreign competition, ultimately weakening defense capability. Too little defense invites invasion. When the emergy basis for a country decreases, needs and budgets for defense are less. Formerly when defense budgets were larger, almost half went into defense research with many innovations spinning off to industry. Now budgets for developing new technologies are smaller.

One lesson from World War II is that a policy of isolation allowed totalitarian regimes to develop a pattern of unopposed conquest, which eventually caused all the nations huge losses in wars of survival. Therefore, some of the world emergy budget has to go to prevent disturbances to peace (see Chapter 9). Mutual aid and peacekeeping missions shared with other nations require fewer resources than a go-it-alone policy. So long as nations at the top of the global energy hierarchy—like the United States—are receiving more net emergy in their trade with other countries, they have the responsibility to do more of the Earth's peacekeeping.

Allowing sales of arms abroad is too great a risk, regardless of the need to balance dollars in foreign trade. The emergy of money earned from arms sales is much less than that required for military force to control terrorist use of those arms.

Drug Traffic

The international flow of money for illegal drugs is so large that major corruption impacts some nations (Columbia and Mexico) and many neighborhoods in the United States. The U.S. efforts to stop drug manufacture in these countries hurts foreign relations. Addictive drugs such as cocaine and tobacco have very high transformities, which means they have huge effects, as everyone knows.

If the data can be obtained, an emergy evaluation could determine whether an emphasis on law enforcement is a net benefit. The study should evaluate a legalization alternative where funds–now used to fight drug transport and sale–are used for education and treatment. The history of alcohol prohibition in the 1920s may be comparable. After alcohol prohibition was lifted, crime and corruption decreased and taxes were obtained, but alcoholism has continued to be a costly problem requiring education, treatment, and prevention of drunk driving.

CHARACTERISTICS OF NATIONAL CLIMAX

The crest of a nation's development is sweet and sour with its accomplishments, information storms, unbearable complexity, accelerating change, crowding of individuals, and stress on paradigms of belief. But people can be happier if they can understand and revel in the amazing turbulence of transition. Shared vision on national goals is needed, but not to the degree that rigid laws inhibit self-organization for new stages.

Role as High Technology Center

Some people believe that the way the United States (or any nation) can maintain its economic leadership is to become the world's information center and balance its trade deficit with a steady output of new products (computer hardware, software, innovative communication, literature, art, movies, music, and celebrity endorsements). However, already many other countries are sharing the role as world information centers. Because information and technology have very high transformities (see Chapters 1 and 4), they are easily transmitted and become widely shared. It is hard to collect money for these or keep them from becoming part of the free global information pool. For example, most of the U.S. innovations and manufacturing in television went overseas. Future innovations will be less if education at home and school falters and research budgets in industry and federal and state governments are cut further to minimize taxes and maximize short-range profits.

Diversion into Sports

Apparently societies with competition-oriented cultures turn to sports when their economies crest, as an outlet for competitive feelings, as a diversion when opportunities are few, and as a way to maintain social coherence. The social satisfaction of *winning* as a group or supporting a winning team may be needed when competition is no longer so useful in the workplaces or in military adventures. Nearly everyone enjoys having a topic that people of all economic and cultural levels can share. But recent years' excess waste in big-time sports detracts from the sport and depletes society. Putting a limit on costs can bring common sense back to sports without detracting from its role in unifying society (see Chapter 12). Emphasis on sports teams, now distorting university purposes, could be managed through independent sports clubs instead (see pp. 262–266).

Higher Selfishness of National Contribution

Ecosystem analogies help us see that it is good national policy to help maximize performance of the larger global system. Except in early stages of colonization, cooperation among units maximizes empower (see Chapter 5). Policies to improve global functions increase the welfare of component nations. Keeping the climate stable helps sustain the economy. Making a trading partner prosperous assures continued jobs at home. Sharing military expenditures reduces the costs of defense for each nation.

SUMMARY

Developed nations can maintain what is essential in their way of life for some time as long as the fuels and other imports necessary to their emergy budgets can be paid for with a favorable international balance of trade. This can be achieved by energy conservation, especially by decreasing the wasted emergy in the unnecessary use of automobiles and fuel. Trade treaties can moderate inequities of free trade. Military costs can be shared more with other countries responsible for peacekeeping and making fuel available globally.

Changed attitudes can foster a more cooperative society no longer expanding, analogous in principle to the mature ecological system. New policies should emphasize information innovation, efficiency rather than speed, cooperation rather than competition, diversity rather than conformity, good maintenance rather than growth, and suppression of borrowing. A national condition with more security for individuals can be sustained by replacing luxury and waste, as explained in the next chapter.

12

SUSTAINING PEOPLE

On the scale of human life a mature civilization keeps everyone at work, sustains health, and educates children. An appropriate distribution of wealth, power, and incentive is necessary for people to be productive and prosperous. In this chapter we use systems principles to suggest effective roles for humans for the time of transition and climax.

At maturity a system changes its emphasis from growth to efficiency (see p. 84). During the time of leveling population, only one or two children are appropriate for a family. For these times, family patterns organize to help each adult do satisfying work and each child to *learn the pleasure of learning how to learn.* For maximum productivity, institutions of health care help sustain good health. A distribution of wealth is needed that favors everyone's work.

Coupling of Consumption and Production

In a mature nation, just as in a climax ecosystem, production and consumption tend to balance and reinforce one another. In the human economy, people are both consumers and producers. In an efficient economy, people consume whatever helps them to produce. Good use feeds products and services back to increase production (Figure 12.1a). Poor use diverts products into luxury and waste (Figure 12.1b). Although both arrangements circulate money and make jobs, the reinforcing feedback maximizes intake and efficiency whereas a system with waste and luxury (Figure 12.1b) does not.

> Jobs that circulate money into waste and luxury do not sustain the economy as much as jobs that reinforce productivity.

Figure 12.1. Comparison of useful and wasteful designs for connecting production and consumption. Pathways with solid lines are flows of emergy; pathways with dashed lines represent the circulation of money. (a) Symbiotic reinforcement of production and consumption; (b) reinforcement weakened by luxury and waste.

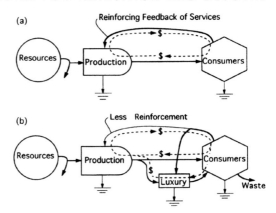

Productivity is achieved by keeping people healthy, employed in useful work, and with functional home lives. Sustainability requires effective childcare and education for the next generation.

DISTRIBUTION OF WEALTH

For two centuries of growth, accumulation of capital was necessary to make the economy competitive and maximize its prosperity. The growth also concentrated economic wealth in a few people. Adulation and admiration of the wealthy and their excesses might have been appropriate images of a society performing net growth. But with the shift from growth to efficiency in our period of transition and climax, a different ideal with less respect for luxuriant waste is expected. During the transition we can expect views to change:

- The majority had faith in a free market paying whatever wage was competitive. For efficiency, a policy is needed that pays whatever wages sustain human productivity.

- The majority had faith in the right of individuals with money to make money without work (unearned income from stocks and bonds). At climax, unearned money needs to sustain production.

Hierarchy of Wealth and Power

Wealth and power produced by a nation are distributed unevenly among its people. Recall the pattern of hierarchy that is natural and observed in systems generally (Figure 4.8). Curve #1 in Figure 12.2a is the shape expected for the hierarchy of people. The majority of people at a lower level support the few people at a higher level. This distribution reflects the natural energy hierarchy of self-organization (see Chapter 4). The transformity of a person's occupation determines the position from left to right in Figure 12.2a. Higher empower goes into

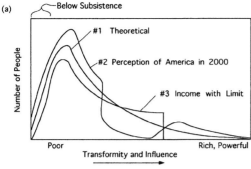

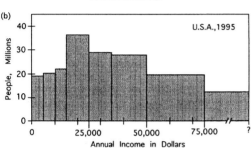

Figure 12.2. Distribution of people in the hierarchy of wealth and power. (a) Three concepts of the way people are distributed on the scale of transformity, a measure of individual worth and power; (b) distribution of people according to their income in the United States in 1995.

and is exerted by political leaders, celebrities, and investors (on the right). The public perception in 2000 was of a society with many people too poor and a few too rich (curve #2 in Figure 12.2a).

Figure 12.2b shows the distribution of income in the United States in 1995. The zone near zero on the left is below the level of subsistence and had few people. Many people had moderate incomes and a few at the right were wealthy. People regarded income differences as a necessary incentive and money as a measure of success.

Over their lifetimes, people move up the scale of size, time, and worth. They start with the small territory and rapid processes of babyhood, increasing their territory of support and influence as they pass into school, the work force, and senior status, where their knowledge and experience are valued. In other words, a person moves from left to right along the energy hierarchy graph, gaining in transformity (Figures 4.8 and 12.2). Usually, incomes increase also.

However, many people in the United States in the late twentieth century viewed the greater wealth of people in higher roles as personal rewards of no concern to others. Some chief executive officers of large companies were allowed to pay themselves very high salaries and bonuses, a nonreinforcing societal waste. In 1996 the annual paycheck of the average chief executive of U.S. companies had increased four times faster than that of the average production worker. But

FRANK & ERNEST by Bob Thaves. © Distributed by United Feature Syndicate. Reprinted by permission.

excessive personal salaries were argued in public controversy as a pathological distribution of wealth.

The emerging public attitude was more in keeping with the service ideal in which more efforts to reinforce society would be expected from those receiving more income. It was regarded as appropriate for a person with greater responsibility to have a somewhat higher income, but not so that money was diverted to luxury. The belief that participation and incomes should be organized to maximize the system's performance is the *maximum empower ethic*.

Investment and Unearned Income

When there are resources to be developed, a business can borrow money, issue stock, and expand. If successful, there is enough new profit to pay off the loan with interest. Those with money provide the loans and receive interest or stock dividends. Thus, people with money make money without doing work for it. By our definition, *unearned income is money received without work.*

Some capital gains are unearned income also. For example, if city development increases property selling prices, the profit from the land is unearned because it was not from the owner's work.

Most people receive some unearned income. For example, money that people save in banks is used to finance expansion of homes and businesses. The depositors are paid some of the interest money. More and more people are investing in stock markets. When there is expansion of the stock market (a bull market), the stocks can be bought and sold later for a profit.

Borrowing and capital accumulation are appropriate for the growth stage if the unearned income is reinvested in developing more productivity. However, individualistic attitudes often regard the earnings on investments as personal profit, to be used on luxury. Unless the earnings support useful projects or are reinvested, the society is not reinforced. Individual priorities should be consistent with public needs.

Nonprofit Enterprise

Ideas about private enterprise are linked in most minds with the profit motive. Profits require expansion of resources and thus are appropriate during growth periods. Nonprofit enterprises are appropriate when growth is not possible. A business that borrows and tries to grow but can't may lose out to a nonprofit enterprise because of its interest liabilities.

Profit and nonprofit enterprises can be public or private. The political ideology for privatization often implies profit making and growth. But for a nongrowth transition, nonprofit enterprises are appropriate (public and private).

Universal Limit to Individual Income

Policies to limit income are already under discussion (see Chapter 3). A nation could place a universal income limit on individuals. For example, a maximum could be set at $150,000 (in 2000 dollars adjusted each year for inflation). The Internal Revenue Service could collect 100 percent of income above that limit. This policy is appropriate for an economy at the summit when large-scale profit is not needed. It could eliminate some of the differences between rich and poor in U.S. society. Quality of work might replace income for judging merit. The range of income would be less (alternative #3 in Figure 12.2a).

The policy on individual income *should not apply to business earnings*. Since excess profits could not be paid to individuals, business earnings would be reinvested or used to support worthy causes.

The American Dream

In the United States, journalists like to write about "the American Dream," sometimes described as "the opportunity to progress to the top as part of expansion and growth." Hopefully, more people will pursue a revised national dream for the times of transition:

> A good living, a feeling of social justice, pleasure in doing a good job, progress in small steps, and satisfaction in seeing a few children grow (yours and others).

EMPLOYMENT

An efficient nation has all its people employed in useful work contributing to production. Whereas private enterprises during growth could absorb new workers, a mature nation without growth may need special mechanisms to sustain full employment.

Like the fast-growing weeds of a newly colonized ecosystem, developed countries in their growth stages experienced exponential growth, competition, and monopolies (analogous to monocultures).

In the early stages, borrowing was important to finance exploitation. Borrowing allowed businesses to make faster starts. Other things being equal, the first competitor to start exponential growth won the competition. Those who lost jobs when a competing business was eliminated were able to transfer to the winning business, which was expanding and employing more people.

Permanent Public Works and Universal Employment

A solution to unemployment might be to restore permanently the program of public works that was successful in the Depression of the 1930s. Jobs would be provided to all the unemployed at minimum wage with no time limit, although many would move on to higher paying jobs in the private sector. The program would be likely to pay for itself not only in the work done, but also in the savings in unemployment insurance, welfare, crime, law enforcement, health care, and education. Youngsters, including teenage mothers, can find their first job in this program. Jobs could include the environmental works in parks like that done by the old CCC (Civilian Conservation Corp), restoration of ghettos, aides to teachers, aides in hospitals and hospices, clerical aides in government offices, aides in research programs, and aides to the handicapped. With nearly everyone employed, many poor would gain self-respect; there would be fewer idle youth; fear of the streets would diminish; and a spirit of mutual helpfulness suitable for a mature nation could replace the selfish competitiveness of the growth times now passing.

Minimum Wage

During growth, minimum-wage jobs were regarded as temporary occupations for many workers who would later move to better jobs. But in a nation without much growth, some people may have to work at the minimum wage permanently. Hence a decision on the size of the minimum wage becomes a decision on what is every individual's minimum share of the national wealth based on emergy. For national efficiency the question should be, "What minimum wage makes the individual self-supporting and capable of doing his or her part in each citizen's unpaid duties such as child rearing, maintaining satisfactory sanitation, and supporting the neighborhood?"

Sliding Wage Scale with Age

To encourage employers to hire youngsters, getting them off the streets and into the workforce, a sliding scale for the minimum wage would be appropriate. This sliding scale, suggested by former President Ronald Reagan, would contribute to full employment and national efficiency. Perhaps $.50 might be subtracted from the minimum wage for each year of age less than eighteen. For those of school age, employment could be outside of school hours, part-time only. A wage scale by age can be viewed as an apprentice program.

Reduced Wage Rate and Part-Time Work for the Elderly

In an efficient nation all its people need to contribute to production. Automatic retirement practices take people out of the work force and require huge costs for pensions and social security. Although many older workers' individual energy, endurance, and ability to learn may have decreased, their knowledge, experience, and stability are valuable assets.

A more efficient policy might replace full retirement with part-time employment at a lower wage rate subject to health limitations. For example, a school system could allow teachers at age sixty-five to go to half time at a lower hourly rate. Social security could supplement salaries with a flat rate that increases with age. This policy of delaying full retirement should help the crisis in funding social security caused by the increasing percentage of older people expected in the next few years.

Employment in Information Technology

Many propose to expand employment and pay rates by stimulating more high technology computer industries and information processing. The idea is for a nation's economy to become the world's high technology center. But no nation can be the world's information monopoly, since more and more nations have developed their own information industries and centers. Nor can an economy live on information alone (see Chapter 13). However, computers are permeating all sectors of all economies, becoming an efficient tool for other kinds of jobs and joining the nations of the world in common enterprises. It should be possible to sustain these ways in the transition-climax.

ECONOMIC ENRICHMENT BY REDUCING CONSUMPTION

Because the money and emergy flows of our economy and its foreign exchange are so interdependent, sustaining jobs against foreign competition depends on eliminating luxury and waste. Emergy that is saved can go for important needs. With less waste, exports are cheaper and compete better; wages buy more. The general principle is:

> Wasteful indulging reduces more important enterprises.

Excess cars and unnecessary horsepower are a major waste in U.S. society. With two diagrams in Figure 12.3 we show how an excess of cars diminishes job availability. In the simpler diagram above, the emergy of the home resources flowing in from the left supports labor and generates export sales. The money from sales buys imported fuel emergy.

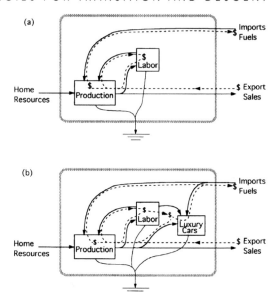

Figure 12.3. Systems overview of a national economy, its resource basis, and foreign exchange. Flows of emergy (empower) are solid lines, and money flows are dashed lines. (a) Without luxury and waste; (b) with luxury cars and their excessive waste. Luxury cars are shown without reinforcing feedback.

In the lower diagram, the luxury cars added divert and consume the emergy of home resources and imported fuels. Less emergy is available for export sales, which means fewer jobs and less money to buy imported emergy. The result is a lowered standard of living. Conversely, reducing the number of cars and their horsepower can sustain an economy, its balance of trade, and jobs. This is especially important at the summit when the emergy yield ratios of fuels decline. Of course, the process of reducing the number of cars also reorganizes transportation.

People raised with an emphasis on individual rights can have a blind spot, not seeing that one individual's waste hurts others. Nearly everyone underestimates the impact of wasting fuel, because its effect on the economy is six times larger than the money involved (when the net emergy yield ratio is 6 to 1–see Chapter 10). But during transition, attitudes may be expected to change.

HEALTH CARE

Originally, human populations were partly regulated by diseases, which provided a death system for weakened individuals often before old age. With progress in medicine over the last century of growth, many diseases were defeated, at least temporarily. Now people live longer and die of body malfunctions such as heart failure and cancer (failures of the immune system). With more and more medical technology it has become possible to keep people alive longer and longer, even after their ability to carry out simple services is lost. The medical ethic of our culture (taught in many churches and medical schools and written into law) has been to do whatever is required without limit. But sustaining longevity is expensive.

As the average age increases in more and more developed countries, the increasing health care costs of the old medical ethic begin to affect the national economy. Countries with younger populations and less health care and sickness can export at lower prices, competing better in international markets. (This is another example of the way large internal emergy diversion can affect a nation's economy.) As former President Bill Clinton explained, increasing health care costs are limiting the strength of the United States in other sectors. The United States is spending more on health care per person than other developed nations, but is providing health care to a smaller percentage of its citizens. A new health care system is needed.

Infant Mortality and People with Disabilities

Originally, a high infant mortality rate removed babies susceptible to disease and those with weak constitutions or birth defects. Families a century ago had many children, several of whom died as infants. Many deformed fetuses were automatically eliminated by miscarriages, thus allowing families to replace the genetic errors with another child.

Advances in medicine and health care made it possible to carry more pregnancies to term and keep more malformed fetuses and newborns alive, sometimes at great cost to families and society. Without normal selection, more and more birth defects are inherited and appear again in the next generation. There is also an increase in damaging chemicals affecting pregnancy, such as drugs, alcohol, and pollution. The high energy of cities, cars, and industry disables more people in accidents. In modern civilization it is good policy to help all to contribute as much as they can. Many special education programs have been developed; physical infrastructure has been adapted to wheelchairs, and so forth. For the lower-energy times ahead, when society can't provide as many resources for special programs, the natural selection against defects is likely to increase. Humane ways are needed to prevent people from becoming incapacitated in the first place.

In past societies with limited resources, various means of population limitation now regarded as unacceptable—such as sacrifices and infanticide—were employed.[1] As the public recognizes the seriousness of overpopulation in a time of declining resources, there should be an opportunity to design better approaches to population limitation than those used by the societies of the past. Birth control and early abortion are likely to become accepted.

Ending Life

Public discourse in the 1990s agonized over alternatives for ending life. At an earlier time, contagious diseases often provided a test of an individual's stamina to survive that did not require human decisions. However, when these diseases went into an epidemic mode, young and healthy people died also. The medical ethic in pure form tries to keep people alive even when they are in terminal organic illness, wish to die, or are supported at great cost in a coma with life-substitution systems. People in our present culture don't want humans making life-and-death decisions, although it is sometimes required of judges. Public attitudes have softened toward physician-assisted suicide.

Health Care Priorities

On an impersonal, large-scale view, the emergy allocated to save lives might be related to patients' potential for contribution. But there is not now a practical way to evaluate individuals or their transformities. The present health care insurance system provides more care to those with more money. However, the distribution of money among people is often not in proportion to the importance of their work.

Health Management

In the 1990s, health management organizations were expanding to meet the crisis of greater costs for health care while ever fewer people were covered by health insurance. Efforts to economize on health care were not keeping up with surging medical costs. For-profit organizations rushed to fill the need as if it were a new resource in which to invest. Profit and nonprofit health care organizations competed for business. In the past, nonprofit health care organizations (public and private) provided quality care efficiently. Adding profit to health care is an unnecessary diversion of emergy. Instances of abuse were publicized, including cases of decreasing care to reduce costs, dropping sick people from coverage, and raising prices to make a profit. Bankruptcies left people without coverage. The system was in shambles. A system that leaves millions uncovered with no health care insurance undermines national productivity.

A Universal, Baseline Health Care

A health care solution that is consistent with hierarchical concepts and democracy is to set up a nonprofit health care system that provides some basics *for all citizens* and includes pharmaceuticals, simpler operations, emergency treatments, pregnancy care, and preventative medicine such as check-up examinations and vaccines. Long hospitalizations and the more expensive, experimental, or luxury treatments could be omitted. These might be covered as now with private money and medical insurance.

The emergy going into poor health and health care will strain the economy and force cultural change, especially while the population is aging. Until the population decreases again, there will be epidemic diseases, such as those that have begun to reappear in disorganized areas of Africa. Antibiotic-resistant strains of infectious diseases, such as tuberculosis, gonorrhea, and streptococcus, may spread. Some of the advantages of a half century of pharmaceutical research are apparently temporary. The old system may return in which diseases select survivors by testing vitality. Our humanitarian ethic may be revised to expect less health care.

SUMMARY

Patterns of human life may be expected to change during the time of transition-climax when resources become scarce. Prosperity requires improved efficiency for essentials and elimination of what is unimportant. Productive jobs can replace jobs that supply luxury. Placing a ceiling on personal income can reduce upper-class luxury, waste, and unearned income. A public works program at an increased minimum wage can insure full employment, increase productivity, and reduce costs of welfare and crime. Part-time work of elders at reduced wage rates can replace the system of early retirement, which will also protect the trust funds of the social security system. Various measures are suggested to prevent disabilities and keep everyone productive. A universal, public, baseline health care system is needed for all citizens, leaving more expensive medicine to be covered by private insurance.

The public needs to recognize how the tradition of free consumption hurts the economy by wasting emergy. Collective love of the automobile already limits what people really care about most: their jobs, health, children's education, and environmental quality. By increasing efficiency with a lower-emergy lifestyle, especially by reducing the use of cars to match the rising costs of fuels from abroad, we can sustain for some time what is most important in mature nations. These goals are reasonable for less-developed countries also.

13

STARTING DOWN

Soon the United States and other developed nations will begin their descent, learning how to live with less emergy and a smaller economy. There is no modern experience in coming down to go by, but we do have some principles about cycles from Chapter 5 and the historical record of past civilizations (see Chapter 3). We get some ideas observing ecosystems when they contract. In this chapter we suggest policies to start the journey down.

In 2000, many national systems were, judging by their fuel consumption, still at full power. Innovations were continuing in science, communications, and medicine. But with downsizing of jobs and wages, some affluent consumption was already disappearing for some people. Many authors warned of malfunctions, and some predicted turndown (see Chapter 3). Our energy review (see Chapter 10) left little doubt that turndown will occur. Instead of denial, it is time for people at all levels of society to plan for a better world in which we use less. There should be task forces throughout society working on descent.

CHANGING THE CADENCE

Conscious reform can't really begin until *descent* grasps global attention. The start-of-a-century feelings of confusion may crystallize at any time into a shared recognition of turndown. One reason the human species dominates the Earth is its social psychological ability to adopt those group mores that work. Growth behavior worked while resources were abundant and succession was in a growth stage because that regime maximized empower (see Chapter 5). After several centuries of frenzied growth, most of the world's people believe strongly that growth is good and descent is bad. Human language defines growth words with positive connotations. In contrast, most descent-related

Table 13.1—Downcycle Words from a Thesaurus

Abate	Deploy	Plunge	Secede
Collapse	Descend	Recall	Settle
Contract	Deteriorate	Recant	Shrink
Crash	Devolve	Recede	Shrivel
Cut	Dip	Recess	Sink
Decay	Dissolve	Reduce	Slack
Decession	Down	Relax	Sleep
Decline	Drop	Remove	Slide
Decompose	Droop	Repeal	Slip
Decrease	Dwindle	Replace	Slump
Dediferentiate	Ebb	Reset	Soften
Degenerate	Enfeeble	Rest	Subside
Degrow	Fade	Retire	Subtract
Demassify	Fail	Retreat	Unwind
Depress	Fall	Retrench	Weaken
Dissolve	Lessen	Retrofit	Withdraw
Diminish	Lower	Sag	

words carry a negative feeling, showing our antidescent attitudes. Test yourself by reading Table 13.1.

Some believe that *Homo sapiens* cannot make the major cultural shift necessary to adapt to descent, but the fast changes in social consensus that we see daily suggest otherwise. Global television facilitates the rapid resetting of public attitudes caused by people's shared experiences. Within the time of one generation, we could observe a reversal of attitude so that descent becomes good and twentieth-century growth ideas become bad. As turndown progresses, people will see what works when resources are scarce.

Coming down *doesn't* mean going back to ways of the past. In general, descent means new ways. The new direction means fewer children, more women working for salary, and less borrowing–the opposite of the recent growth society. In a sense, descent can be a kind of progress. Less new technology will come from research institutions with reduced budgets, but many emergy-conserving innovations will be new.

Let's help change attitudes now to avoid the inefficiency of trial and error. The education system ought to teach the changes necessary for a prosperous way down. At some point ahead a political leader, spiritual visionary, or perhaps some beloved celebrity will find the opportunity in center stage to galvanize global excitement to welcome descent.[1]

REDUCING POPULATION

Population principles for climax and descent were explained in Chapter 8. To maintain the standard of living, the population has to decrease at the same or greater rate as the empower. When resources are scarce, and the population is constant or increasing, there are fewer resources per person, and a lower emergy for each.

Programs to limit births have been opposed by many religious and political groups such as those of Christianity, Islam, and early Chinese communism. Religious dogma encouraged large families and gave reproduction a moral backing that was appropriate for those growth periods, especially where death rates from diseases were high. Higher population meant higher empower production.

Now, however, excess populations are no longer good for production because other factors are limiting (Figure 8.2). Population increase divides emergy among a greater number of mouths. More and more of the economy has to go into basic necessities for the increasing population and less can go into production, information, research, and innovation.

ECONOMIC TRANSFORMATION

Policies to keep an economy prosperous are likely to change during the downcycle stage (see Chapters 6 and 11). Economic concepts of earlier centuries may replace neoclassical beliefs because they are more appropriate when there are resource limits.

Inflation and the Ratio of Emergy to Money

During the growth period, those controlling national money policy found that adding some new money each year accelerated real growth. The money supply circulating was allowed to increase slightly faster than the real wealth. One way to do that was to lower interest rates for loans. Whenever there is a loan, new money is created (since the money loaned and the paper promising to repay are both forms of money). Borrowing thus accelerated development when there were resources available to exploit. The competing business with the start-up capital could displace competitors that had no means to accelerate quickly. Thus, during growth, money increased faster than the emergy and the emergy-money ratio decreased a little each year–a measure of inflation.

During descent, however, real wealth (annual empower) decreases due to resource limits and rising prices. Adding money to the circulation when wealth is declining can't stimulate growth and only causes high rates of inflation. An example of this "stagflation"–inflation without growth–was the inflation that followed oil shortages from 1974 to 1984.

With less wealth to process, the money supply needs to be decreased to avoid inflation. Central governments can try to reduce money supply and decrease borrowing by increasing interest rates. The Federal Reserve Bank of the United States, by buying and selling securities to banks, can manipulate the money supply downward. This policy could keep the emergy-money ratio stable and maintain the value of the dollar. A declining stock market also reduces the money supply, a process called *deflation*. With less borrowing, interest rates drop.

People who don't understand the changes might keep trying to expand, borrow, and invest in growth unsuccessfully. This would only generate defaulted loans and bankruptcies. If deflation of money was faster than the decrease in empower, the emergy-money ratio could increase. But severe deflation, in which everyone loses money, could cause a depression, a condition like that in the 1930s United States, when not enough money was circulating to sustain productivity.

Governments of nations that have been on the short end of the emergy trade balance can enact legislation to encourage companies to keep more resources in use within the country. In order to establish more empower at home, policies should stop commodity exports except those necessary to generate the dollars for imported fuels and information (see Chapter 9). Other outflows of money should be limited.

Borrowing and Interest

When resources do not permit growth, neither production nor markets can expand. Borrowing isn't profitable when there is nothing new to produce, sell, and pay the interest. What borrowing there is may be for local developments to improve existing functions. If there is more money for loans than borrowers for it, interest rates will be low.

When growth of real wealth is limited, new money cannot be added without losing its value. If profits and money cannot accumulate, capitalism can no longer dominate. In biblical and medieval times when economies were not growing, borrowing and the charging of high interest rates—called *usury*—were not considered in the public interest and regarded as evil.

Banks

During times of descent, not many new resources are available to exploit. With less borrowing, banks have less business. With underutilized money and low interest rates, banks earn less. Loans made in a period without growth are hard to repay. There are more business failures, bankruptcies, and bank closings.

During the time of descent, banks may acquire new roles such as financing contraction. Some capital will be needed to finance more efficient, smaller, and lower-energy enterprises. When populations decrease or people give up second homes, banks can buy the excess housing and sell to first-time home buyers.

Resource Use and Discount Rate

According to the discount idea, more money is obtained by harvesting and selling resources now rather than later. The extra money comes from the interest paid when the harvest money is promptly reinvested (for example, deposited in a bank account that pays interest). The earning rate from selling now is the *discount rate*. However, the growth processes of nature may increase monetary value faster than the bank's interest payments.

Especially in a time of descent, the real wealth system can hold its monetary value better than bank savings. Environmental systems such as forestry, fisheries, and agriculture generate new wealth continuously. If increases in emergy and dollar values by these production processes are faster than those from exploitation and interest, the discount rate is negative. Then it pays to buy real wealth rather than save money in the bank.

Bank deposits may also lose value from inflation. Inflation is to be expected if the money in the economy is not reduced as fast as the empower.

Stocks

In developed societies, corporations sell stocks to pay for expansion, and some of the profits are paid to stockholders as dividends. Without growth, many stocks can't pay dividends and lose their market value. Declining sale prices of stocks is called a bear market. Advantages of owning stocks decrease. If public perception of descent comes suddenly, there is a danger of a stock market crash. The disappearance of money with sudden deflation would make it difficult to pay for the reorganization costs of descent. A mechanism is needed to program a gradual, noncatastrophic deflation of the excess money held in stocks and bonds. For example, an economy-wide, step-wise limitation of interest and dividend rates could shift money from stocks and bonds to ownership of operations adapted for production during descent.

Stocks of companies emphasizing the reorganization and contraction of society could rise in value temporarily. Possible examples are companies that provide remodeling, repair, basic medical supplies, and farming tools. Municipal bonds may be issued to make utilities more efficient. But in a declining economy even these have only a small growth potential. In these conditions stocks may be more appropriate for small earnings and less for speculation.

Reducing Salaries and Wages

Perhaps the most important descent policy is *cutting salaries uniformly*. Suppose reduced volume of business requires a 10 percent reduction in salary and wages. Instead of firing 10 percent of the people, cut everyone's salary 10 percent, keeping everyone employed. This policy is equitable and democratic, because all persons are free to determine what part of their consumption is personally less important. Full employment keeps money circulating, decreasing unemployment insurance, family stress, welfare, crime, and the high cost of jails that might otherwise result.

With salary reduction, each company can reduce costs and at the same time keep its trained workforce. Productivity is sustained. Keeping people at work gives more security, gains loyalty, and holds teams together. Flexibility in hours worked allows more efficient fitting of work and home responsibilities. After wage reduction, workers still have the option of going to any available higher-paying job.

Compare two examples: In the Texas panhandle towns of Pampa and Borger, where dozens of oil rigs had shut down, public school-teachers received contracts that called for salary cuts the next year of $1,000 to $3,000. In Oklahoma, where low oil prices had devastated the state's tax base, about 10,000 teachers were given notices that they would not be hired the next year. In which state would you rather be a teacher?

By trying to hold on to wages that don't compete internationally, unions can destroy their own jobs. In times of little profit, pressuring for raises forces companies to contract tasks out to cheaper, nonunion workers or to move the industry to a place where labor is cheaper. If the company does not survive, neither will union jobs. For example, Eastern Airlines went bankrupt after failing to persuade unions to cut wages.

Instead, unions could help control necessary across-the-board wage cuts, thus increasing job security and continuing full employment. They can insist that downsizing cuts be at all levels of management as well.

International Minimum Wage

To limit the way global capitalism can exploit labor by shifting operations from one country to another, we propose an international minimum wage, agreed to by treaty. Wage rates can be made similar by adjusting the minimum in each country to have the same real-wealth buying power (*emergy accounting*–see Chapter 9). Instead of pulling *down* labor of developed countries, it will bring labor of underdeveloped areas *up* to a level that can sustain global production.

Competing at Conserving

Our cultural tendencies to compete may show up as competition for downsizing opportunities.[2] Already there has been some bidding for management reorganization of state government functions in Florida. Austerity proposals need to be nonprofit to be competitive. Church- and municipal-operated hospitals and clinics are examples of nonprofit private enterprises. Unlike the competition during growth, competition for an efficiency role is not an exponentially increasing process that overgrows and displaces diversity.

Emergy Investment Ratio

In Chapter 6, the emergy investment ratio—the ratio of purchased emergy to the emergy from the local free environment—was introduced for evaluating the economic inputs to environmental use (Figure 6.6). With less fuel in use, this ratio diminishes. A low ratio indicates that there is less pressure on the environment for matching (for example, less pressure to use lands for developments). With few purchased inputs, the economy makes products with lower prices and is more competitive. On the other hand, the environment becomes a higher proportion of the total economy (for example, people returning to rural land occupations). By estimating what the investment ratio of an area will be as it declines, the appropriate intensity of industries and environmental uses can be planned.

RELAXED HIERARCHY

Recall from Chapters 4 and 5 the hierarchical principle believed to be universal (Figures 4.8, 5.6, and 12.2), that many activities of smaller

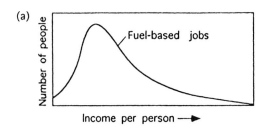

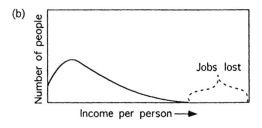

Figure 13.1. Change in the distribution of people and jobs during descent. (a) Long hierarchy based on renewable and non-renewable resources. Also see Figure 12.2; (b) shorter hierarchy based on renewable resources in the future.

energy are required to support a few of higher-quality energy (higher transformity). During descent, the resource base will be smaller with less total empower, causing the hierarchy to shorten, with loss of components at the top. As the fuels and total empower diminish, the curve of wealth distribution (Figure 13.1a) is decreased (Figure 13.1b). Job hierarchy shrinks, leaving fewer positions at the top. The trend toward fewer administrators was already being noticed in the year 2000.

Similarly, as Figure 13.1b shows, society will be unable to support many of its very special high-transformity systems. We may expect reductions in research, space, sports extravaganzas, military super-weapons, super communications, university size, and celebrity worship. In areas of high risk when natural disasters destroy settlements, replacements may not be necessary.

Government Role

The shorter energy hierarchy would mean less government. However, the necessary simplification of government roles that goes with a lower resource base should not be confused with ideologies that advocate no government control, unbridled competition, wild economic speculations, unsafe banking, unmonitored airlines, or citizens free from law. One danger to prosperous descent is the loss of societal cohesion now provided by government functions. Another is the risky, growth-seeking effort of uncontrolled global capital investments (see Chapter 9).

Already in 2000, policies had been initiated to reduce government, increase efficiency, and conserve public resources. There were already reductions in subsidies and incentives for births, immigration, roads, mines, and agriculture. Once the major negative role of excess cars is recognized (see Chapter 12), we can expect less use of highway expansion trust funds and depletion allowances for oil companies.

The present global hierarchy converges emergy and money to centers in a few nations. For example, centers of the world's finance and banking have been in New York City, London, and Tokyo; and high-tech research and development centers have been in Silicon Valley, central Texas, and the Boston area. As economies level and equalize, the trend to shorter hierarchy should lead to decentralization and the development of more and smaller centers. Austerity may be expected in present centers of commerce and knowledge as they adapt to less support.

REORGANIZATION OF ENERGY USE

A decrease in available fuels and the resulting higher prices will cause a profound reorganization of human settlements and their use of transportation, which we describe in the next chapter on cities. The auto age will come to an end when alternate needs for the fuels running the personal automobiles become more important than the time saved by having individual cars.

Electric Power

In our civilization, electric power makes the high-transformity emergy age possible. Electric power plants supply a form of energy intermediate in transformity between the fuels and the information systems of commerce, finance, universities, hospitals, government, city management, and residences (see Chapters 6 and 13). When fuels become more costly so does the electricity, because most electricity is made from fossil fuels.

Information centers may be expected to develop more around mountains where hydroelectric power—a renewable resource that costs less because more of the work is done by nature (see Chapter 10)—is more abundant. Hydroelectric plants in the Alps in the center of Europe, for example, already support information centers in Switzerland, Germany, Italy, Austria, and France.

Everywhere measures will be needed to increase efficiency and restrict unproductive consumption (see Chapter 10 and Figure 12.1). Saving money by using more efficient electric appliances is already a public policy. Evaluated on an emergy basis, these conservation measures have even larger contributions to the economy (see Chapter 10). Former President Jimmy Carter, ahead of his time, called energy conservation "the moral equivalent of war."

PRIORITY AND SCHEDULE

Although we don't know enough about descent yet to set priorities and schedules, some thoughts may help.

Information Priorities

Shared information, cultural attitudes, business networks, and common television programming are developing some kind of global unity. As explained in Chapter 9, the spread of global information may be replacing defense organizations as mechanisms to maintain world peace. If the descent is to be prosperous, priority has to go to sustain global information systems even though not everything of high transformity can be retained. Reserving the electric power required for information processing should have priority. With less opportunity to compete for profit, advertising is likely to decrease.

The support for the huge store of knowledge in information institutions (universities, libraries, government centers, private research laboratories) might have to diminish, but much of the knowledge can be organized for long-range storage in order to be available when resources allow other pulses of growth. Without the huge fossil fuel

Credited cartoon by Draper Hill © 1997, The Detroit News.

storages, the next growth cycles may be smaller, based on other environmental accumulations.

More academic attention needs to go into synthesizing, summarizing, and defining essential principles (see Chapter 13). The physical lifetime of library books is short (fifty to 100 years), requiring continual replacement. Whether electronic memory storage can be more reliable remains to be seen. Chapter 17 explains the roles of universities generating ideas for change and serving as society's long-term memory.

Balance and Timing

Main sectors of the economy–those that process and provide food, clothing, housing, health care, recreation, spiritual education, information, infrastructure, and so forth–are necessary to the production and consumption processes of the economy. The economy can be unnecessarily limited if any one sector is either neglected or supported excessively at the expense of the others. Prosperous descent requires a balanced reduction.

How fast do we have to descend? Overview models simulated on computer (see Chapter 5) show how fuel availability can control descent. Reserves of natural gas are not fully known. Coal reserves are large but have lower net yields. Considering them all (see Chapter 10), we conclude that the decrease in net emergy can be gradual.

In our models, the depreciation rate of the stored economic assets also determines how fast the economy descends. Whereas infrastructures such as highways have a replacement time of only fifty years, the information stored in billions of people (culture, religious beliefs, technology, and knowledge) has a slower replacement time. By sharing information globally, we decrease the depreciation rate of what is most important in the civilization. There will be time for gradual orderly change toward an agrarian economy again, if we don't fall apart in the meantime. The descent of the Roman civilization took about 300 years.

APOCALYPSE

An alternative way down is to crash. Nature provides plenty of examples of this, such as the rapid consumption of ecosystems by fire and locusts (see Chapter 5). Archaeological studies of ancient abandoned cities record the rise and decline of earlier civilizations, such as Athens, Persepolis, and Timbuktu. But distinguished authors have many theories as to why (see Chapter 3).

The quickest way to come down, analogous to fire, would be to continue our frenzied resource consumption past the mature climax stage. Continuing the ethic to maximize profit and consumption can generate overpopulation, inflation, shortages, debt, bankruptcy, starvation, riot, local war, frustration, crime, loss of organization, loss of knowledge and libraries, and abandonment of ideals. The environment might lose the last of its reserves (its *natural capital*) of groundwaters, soils, forests, fisheries, minerals, fuel reserves, and biodiversity. Overpopulation can consume reserves, and stored assets rapidly until reaching bottom, with starvation and anarchy along the way.

All this reminds us of the Four Horseman of the Apocalypse (Figure 13.2) who come to destroy humans: death, famine, pestilence, and war. It could be argued that these are the quickest and thus most humanitarian ways to get down for a long period of rebuilding and restoration. But this book proposes that a prosperous, gradual descent is possible and better if we can learn how.

SUMMARY

To be prosperous, descent requires a reduced population, less money, and smaller salaries. Governments and banks will need to help finance initiatives for downsize redevelopment, but the economy will issue fewer stocks and bonds, borrow less, and develop lower interest rates. International treaties can control global capitalism with an international minimum wage.

With less energy and a shortened energy hierarchy, some achievements of the climax civilization must become dormant. Landscapes may

Figure 13.2. "The Four Horsemen of the Apocalypse" trample their victims in Durer's woodcut depicting the end of the world. Death *rides a skeletal nag,* famine *swings his scales,* pestilence *lifts a sword, and* war *aims his arrow. Bequest of Francis Bullard, courtesy Museum of Fine Arts, Boston.*

be expected to organize with fewer cars. Information and medicine can be given priority by allocation of electric power. A deliberate plan for change to avoid apocalypse needs global attention. People will need to understand the changes and share a vision of a less-intensive but better world. Table 13.2 summarizes ways to foster tranquil descent.

Table 13.2—Guidelines for Orderly Descent

Make beneficial descent the collective purpose for this century.
Dedicate television drama, literature, and art to adventures about descent.
Accept a small annual decline in empower use.
Maintain a stable emergy use per person by reducing populations in a
    humanitarian way.
Remove all incentives, dogma, and approval for excess reproduction.
Reduce salaries and wages as necessary to maintain full employment.
Keep the emergy-money ratio stable by adjusting the money in circulation.
Borrow less and reduce expectations of profit from stock markets.
Develop economic incentives for reducing consumption.
Develop public opinion, laws, and taxes to discourage unproductive
    resource use.
Sustain the production of the environment.
Consolidate knowledge for long-term preservation.
Prioritize the communication of concepts of international respect and
    cooperation for global sharing.

14

REORGANIZING CITIES

The great cities of the twentieth century had intense concentrations of people, cars, technology, information, and emergy based on a prodigious consumption of fuels and electricity. Soon, however, with fewer resources, cities will have to decrease and spread out to the rural lands on which they must later depend. Parts of many U.S. cities are operating badly and warped by the emphasis on cars, but beneficial reorganization is possible in the downsizing and decentralization ahead. This chapter uses the principles of spatial organization to anticipate changes in urban civilization.

As we explained in Chapter 7, cities are the functional centers of a regional system that includes subordinate towns, agriculture, and green landscapes. In the times of descent ahead, decentralization will require reorganization of the whole region. At the hierarchical center of landscapes, cities receive inputs that converge from surrounding areas and send back out goods, services, financial investments, wastes, and controls (Figure 7.6). But many of cities' present problems come from attempts to manage those cities as if they were financially separate from their regional system.

The city's roots in the landscape are now overwhelmed by activity running on fuels, electric power, and their technology (Figure 7.7). Civilization all over the world became more similar in the twentieth century because of a common energy basis. But urban shrinkage is expected with descent. When ways of life and livelihood become based once more on the very different environmental conditions in different regions, cities may be expected to regain individuality.

PRESENT CONDITION

In 2000, cities ran mostly on fossil fuels and electric power. They had the most concentrated skyscrapers, the most circulation of money, the most information processing, the highest empower density, and the highest transformity (Figure 7.7). When twentieth-century cities first expanded, most people lived near the center, and in some places–for example, the Asian cities of Taipei, Hong Kong, Tokyo, and Beijing, and the European cities of Stockholm, Copenhagen, Paris, and Vienna– people still do.

Urban people found ways to interact with the rural environment outside the city. People subjected to congestion work better with en- vironmental support. For maximum effect, high-transformity emergy (fuels, human service) needs to interact with abundant lower-quality resources of the outer city, namely the green landscapes. In some cities, people used trains to arrange their urban-rural interactions. In the United States, cheap fuels, expressway construction, and affluence allowed many families to buy homes in the outer fringes–suburban areas–and use personal cars to commute to work. But large paved lots and ga- rages were required to park the cars, often displacing green areas. Pavements increased heat load in the summer.

In some old European cities there were numerous parks where people could interact with green environments. The higher retail prices of fuels in Europe limited cars and helped keep people living inside the city. Most buildings were limited to about six stories with shops at street level and housing above. Sunshine supported shade trees.

Areas of Disorganization

In the United States the flight of people to suburbs was acceler- ated when expressways built without regard to established commu- nity patterns disrupted living areas. The zone between the city center and the belt of affluent residences became disorganized by all the highways and commuter traffic. Buildings were abandoned when in- dustries moved elsewhere for more favorable conditions. When people with more money and better jobs moved out of the inner cities, the areas around the central business district were left with deteriorating apartment houses, a diminishing tax base for schools and services, and failing infrastructure. This inner zone deteriorated with slums, unemployment, poor and homeless people, crime, drugs, fear, and despair. Several authors reviewed in Chapter 3 cited the deterioration of U.S. cities as the most urgent problem for our civilization. Many cities sought creative ways to make disrupted areas functional again, using tax breaks and other incentives for attracting businesses with jobs, assigning government offices, and constructing better apartments.

Beltway Towns

When people in U.S. cities moved to suburbia, shopping centers developed in the new, more distant housing districts, especially at the intersections of the trunk lines with the beltways (Figure 14.1). These

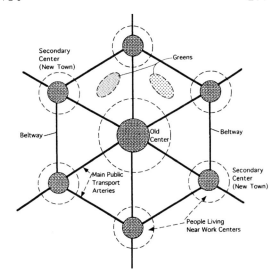

Figure 14.1. Spatial pattern of cities decentralizing into smaller centers on beltways.

shopping centers for the residential suburbs formed a ring around the original city. Examples are the small satellite cities with their own industries, shopping, and housing around Philadelphia, Houston, and Atlanta. Occasionally, special activities like the theater and business at city hall drew people to the old central city, but much of life was in the outer towns. Many industries developed near the secondary centers.

Although the development of secondary centers and towns was partly motivated by the dysfunction of the inner city, the result in Figure 14.1 is consistent with the basic pattern of spatial hierarchy shown in Figure 7.6. Furthermore, the decentralization is an appropriate first step in preparing for the time of descent.

REINTEGRATING CITIES AND REGIONS

To solve present urban problems and to ensure a smooth process of reorganization during descent, cities need to be reintegrated politically and financially with the whole region they serve and from which they need support.

In colonial America, the county (or township) was the regional unit. The courthouse–the center of the county government–was usually located in the largest town and exerted control over the whole county. Within the region were smaller towns that had smaller territories and jurisdictions, subordinate to the central county government. The pattern was functional and fit the principles of spatial organization (see Chapter 7).

But when fossil fuel–based inputs caused the explosive growth of a city, it developed a larger economy than the rest of the region. The emergence of urban concentrations was an example of the development of a new level of hierarchy shown in Figure 5.6. State legislatures allowed city government to become partially independent of the regional government, leaving the county to administer only the areas beyond the city limits. Then when the flight of people from the central cities occurred, the city was left with little resident population or tax base. Many functions such as emergency services and police were duplicated. City and regional governments were often in jurisdictional conflict.

The city and regional managements need to be recombined. *The region is the city's territory and the city is the region's center.* One way is to put both the city and the rest of the county back under one central government, making city government a subordinate department. Then the countywide tax base can be used to make the whole city and its surroundings functional. The city of Jacksonville, Florida, for example, incorporated the entire county, thus substituting city government for the previous two governments. Whether the overall regional government is called a county or a city, the larger territory and tax base are appropriate.

RESTORATION AND DECENTRALIZATION

Most people would agree that the first objective of reintegrated urban and regional initiatives should be to solve the current urban problems. Fortunately, some measures to solve current problems also help prepare for descent. The decentralization that started with suburban housing and beltway towns can be expected to continue. Thus, many actions can improve the city and also prepare it for the future of reduced amount of fuels and electric power.

Cities already face costly replacement of deteriorating roads, buildings, subways, bridges, pipes, electric lines, and other structures. Cheap materials and energy to replace such structures are not so available as when these were built fifty to 100 years ago. Therefore not so much structure can be maintained as before.

Figure 5.6 shows how a landscape with fewer resources loses higher-emergy structures and functions. Some of the highly specialized urban jobs are lost (Figure 13.1b). Cities may be expected to be less concentrated and to diminish in size. As fuels, electric power, goods, and services become more expensive, fewer people can commute. With fewer cars, populations can recluster either around towns further out or around the original center. With people closer to their jobs, restored areas can be efficient.

The information processes that serve the whole region—including government, finance, universities, libraries, computers, and communi-

Courtesy of Randall Enos for The Washington Post.

cation centers—can be retained in the old center. A rebirth of some inner-city living is likely. For example, parts of San Antonio, Texas, Washington, D.C., and Gainesville, Florida, are attracting people to live in the inner city again.

With reorganization, the total activity and wastes are likely to decrease. Later, with less structure and density there is less to maintain, lowering costs and taxes. Less-intensive cities allow more inclusion of a green environment. By anticipating the decentralization, city planners can facilitate changes needed to keep the economy functional.

More clustered living will allow the development of more local social structure than is found in suburban housing scattered widely by commuting. Public works programs for the young and the unemployed can help if concentrated in disturbed zones. With public works reducing unemployment and thus crime, costs of police and jails will decrease. Individuals can spend less on security, guns, multiple locks, alarm systems, and steel gratings. Giant dogs bred for protection can be replaced with smaller pets.

For many parents, the most important factor in selecting a place to live is the quality of schools. Special efforts to spend more on schools in the depressed areas probably have to be made with funds from the next larger scale—county or state.

A pattern of decentralized smaller centers can develop by reorganizing the peripheral shopping centers as new towns (Figure 14.1). When there is less buying power for fuels and cars, people can be comfortable with small business jobs, living in apartment buildings of limited height close to a center without automobiles and parking lots. Two-job families can save time and expense by reducing commuting. Does the family move to a location between the jobs, or next to one job with the other person traveling to distant work?

Strip Development and Advertising

In auto-rich cities, many of the consumer businesses were organized as strip developments along main transportation corridors. These roads were marked with advertising signs, which were often garish and excessive. Electrically lighted signs were both a distraction and waste of high-quality energy. For example, low-priced hydroelectric energy from the government-built Hoover Dam was wasted on lighting the huge billboards of casinos and hotels in Las Vegas, Nevada. Emdollar evaluation could dramatize this diversion of energy from useful production. The efficiency achieved by limiting signs in some cities now can become a general policy during descent. Essential signs of moderate size are needed to identify local businesses and announce new products, but billboards to advertise old products are wasteful, and not appropriate in times of contracting markets and less competition. During growth, extensive advertising created wants, a jump start that accelerated consumer organization. According to the maximum empower theory (efficiency replaces competitive growth—see pp. 83–87), emphasis on expansion and competition may be expected to subside, being replaced with more efficiency, cooperation, and diversity. Strip shopping and advertising may be expected to decrease.

Housing

A fundamental planning error in many U.S. cities was the construction of large, high-density, many-storied apartments and the paving of the surrounding land for cars and parking lots. Many of these buildings were public housing filled with the poor, the unemployed, and those unable to set high community standards. Some buildings were poorly maintained and rampant with drug use and crime. When efforts failed to make these settlements functional, many buildings were torn down.

New welfare plans were tried to avoid concentrating the poor in one area. Vouchers were given to people to rent apartments where there were more jobs and better community structure. New laws required those aided to take part in job training and job placement. Public works programs may be a better way to guarantee jobs and help renovate housing and landscapes for a time of lower density.

As part of increasing efficiency in the current stage of maturity, the building industry is already considering the full "life cycle" of construction and use. The cycle of a building includes initial construction, maintenance during use, environmental exchanges, removal and recycling of materials, and reconstruction. A better building operation uses less energy, reuses materials, lasts longer, and minimizes waste that goes to the environment.

As materials become more expensive, it becomes cheaper to restore an old house than to build or buy a new one. Restoration can be done gradually, and a family can do some of the work. Less borrowing is necessary, and interest is saved. If population decreases faster than the depreciation of housing, an excess of houses could develop, lowering costs of renting or buying.

In Europe, much of the family housing is apartment-sized within permanent stone-and-masonry buildings hundreds of years old. When the interior wears out, the apartments are rebuilt without disturbing the main structure. By comparison, U.S. housing by and large consists of flimsy wood, brick facing, or mobile homes. The turnover time is thirty to fifty years or less. Such structures are appropriate for an early colonization stage of development. We know from examining forest ecosystems the way flimsy weedy structures are appropriate during early growth, but large, durable, more permanent structures such as giant trees become predominant during climax. Such structures with lower turnover time and less depreciation are more efficient in emergy terms. Housing in the United States has been slow to evolve more permanent structures, possibly because of the emphasis on short-term market economics and dispersal in suburbs.

The analogy with ecosystems suggests that weedy, short-life construction used in colonization will be replaced by long-lasting structures. The less-expensive houses built for quick profits will become less important, and more permanent structures will gradually take over. With less transportation and commuting, people may form small clusters. Where buildings are close or touching, public safety regulations require permanent construction to lessen the threat of fires that would burn many blocks. There is a paradox for planners in times of descent. How can enough capital be arranged for more expensive long-range construction when there is less growth? Without capital, such construction may have to be done a little at a time.

City Greens

As city density decreases, there will be more space for trees, parks, and gardens. Greenery can be reestablished. Some concrete can be removed or perforated for water infiltration to assist the growth of larger trees; some can be covered with soil. If levels of toxic pollution are reduced (especially heavy metals), the city could be safe for gardens again. Empty lots can be turned into playgrounds and neighborhood gardens, with plots for families to plant flowers or vegetables, giving young people meaningful chores. Following a pattern in Rome, a neighborhood association that encourages socialization on the local level can control green areas.

There can be space for play fields, parks, ponds, and natural areas and trails. In these ways, more environmental assets are near home, available to people without cars. Wetlands and floodplains can be restored for water quality and flood control. More of the amenities of suburbia can be made available without using cars to commute and shop in strip developments. No-drive shopping and recreation zones can be established.

Some advocate replacing green spaces and parking lots with more housing as a way to reduce commuting and urban sprawl. However, *infilling*, as it is called, is not appropriate for a time of decentralization because it diminishes people's access to green space. Note the hierarchy pattern in Figure 7.6.

TRANSPORTATION

During growth, planners manipulated the transportation infrastructure to encourage and support development. The high net emergy of fuels and electricity helped the growth of an intensely concentrated urban society. At first, most people lived near the center of the city as it developed. Movement between the country and the centers was along public transportation trunk lines–subways, commuter railroads, trolley cars, and highways–arranged like the spokes of a wheel as in Figure 14.1.

During descent, similar transport can be reorganized to return people to a lower emergy pattern. As information technologies become more sophisticated, some communication can be substituted for transportation. Bicycles, motorbikes, small cars, and buses can predominate, since small size, efficiency of operation, fuel efficiency, and short trips allow emergy to go for other essentials. People may be expected to communicate by phone and computer more, make friends nearby, travel less, use public transport more, and fly only rarely on costly airplanes.

Public Transport

City buses and trains work if organized hierarchically;[1] that is, with main carriers on only the major arteries. Public transport can be straight, quick, and frequent if mostly on the main trunk lines and beltways (Figure 14.1). Large buses don't work if their routes meander through residential areas, because too much distance, fuel, and time are used gathering people. People can use smaller equipment such as bicycles, motorbikes, and minibuses to get from their houses to the main trunk lines. No-drive shopping areas accessible from public transport, such as those in Miami, Florida, and Geneva, Switzerland, are already popular.

Bicycles

People living near centers will not need their own cars for daily life. Renting cars for long trips is already becoming a trend. Bicycles will suffice if bikeways are given priority in the organization of roads. Cities in the Netherlands, Austria, and Denmark already have bikeways separated by grass or curbs parallel to the auto-roads. Smoothly paved and broad, with separate stoplights, the bikeways are a practical alternative for transportation to centers from several miles out. Pedestrian walkways are not shared with bikes.

REUSE AND RECYCLE

Successful systems circulate materials without accumulating wastes–a basic principle shown in Figure 7.1. A sustainable system routes by-products into three kinds of recycling in order of decreasing quality:

- *products reused:* usable items such as household appliances are reconditioned (recycled among people)
- *high-quality materials reprocessed:* iron, aluminum, lead, and glass are recycled into manufacturing inputs
- *dilute wastes recycled to nature:* waters containing organic waste particles and chemicals need to be dispersed over a broad area to those ecological systems, such as wetlands, that can use them

Reuse is economical; reprocessing is economical when subsidized; environmental recycling is necessary and a net emergy benefit even when not economical. Transformity, a measure of concentration of value, can be helpful in determining how much effort to spend on a waste item.

Solid Waste Alternatives

U.S. throwaway culture and its accompanying landfill dumps are a terrible waste of emergy, a loss of materials, a diversion of land production, and a source of groundwater pollution. Accumulating solid wastes in landfills hurts the economy. In 1983, Texas wasted 11 percent of its gross state emergy budget on putting municipal waste into landfills.[2] Reusing products, reprocessing materials, and dispersing dilute wastes through wetlands can add to the states's emergy productivity and real wealth.

As resources decrease, reuse and recycling of materials will become more and more economical. Already businesses have developed for reconditioning and reselling appliances. As more mineral resources have become scarce in the United States, it has become economical to collect and reprocess materials such as steel, aluminum, copper, and glass. Some landfills are being dug up again so that valuable metals can be retrieved.

Most cities already pick up selected material for reuse (paper, glass, plastic bottles, aluminum, and lead batteries). When the costs of fuels increase, the cost of plastics (made from petrochemicals) will increase as well. Use of throwaway plastics will decrease, and people can save many of their plastic containers for reuse. Recycling of plastics to plastic manufacture (operating on a small scale in 2000) will develop further. Some public subsidy is justified because the emdollars involved in recycling and environmental protection are greater than the market value of the materials being recycled. For example, a tax on packaging could help.

Requiring more industries to be responsible for the waste from their products could stimulate the reuse of energy-intensive materials. For example, "bottle bills" in some states require payment for the return of drink cans and bottles to retail stores. *Industrial ecology* is a new name used for industrial concepts that take responsibility for efficient management of the full cycle of materials through manufacture, use, and reuse, and environmental dispersal of dilute wastes[3] (see pp. 38–42).

Increasing costs of fuel will make solid waste collection more expensive. There may be incentives for less food packaging, and more bulk delivery to markets where customers can use their own containers. Paper may be used more as prices of plastic increase.

Incineration of solid wastes is wasteful, and difficult to do safely because of toxic fumes from burning plastic. Usable resources are burned unnecessarily, and relatively little electricity is generated. Many components are worth more in reuse than as fuel (higher transformity). Capital investment is costly. Wastes will become less of a problem as households become more efficient and the calorie content of the waste decreases.

After reusable items, concentrated materials, and toxic substances are removed, solid wastes can be shredded and dispersed in forests. W. Smith, experimenting in Florida with a sixteen-inch mulch of municipal wastes, increased growth in planted pines. Adding litter from society to the normal forest litter of falling leaves and branches was beneficial.[4]

Leaf fall and pruned or fallen limbs left dispersed on lands or in local compost piles—instead of concentrated by garbage trucks—can help to restore soils. Collection costs are reduced, and compost becomes available locally for gardens, city shade trees, and wildlife areas.

Wastewaters

The lower density of cities in a time of descent may make it easier to get good natural use for leftover dilute wastes by recycling them through wetlands or into forests. Restored or constructed wetlands not only engage solar energy in waste processing, but also provide wildlife centers for aesthetics and education (see Chapter 15).[5]

Wetlands and fishponds can be established within decentralized cities (Figure 14.1). The drainage basins built in cities to delay storm water runoff can be planted with self-maintaining wetland vegetation instead of mown grass. Wetland retention basins reduce costs, increase water quality, and contribute aesthetic values. If each basin has a small and deeper pool with permanent water, stocks of fish can reduce mosquito problems.

CITIES IN DEVELOPING COUNTRIES

As explained in Chapter 9, overpopulation and the impacts of global economic change on developing countries sent large populations of rural people into their cities. Uneven investment and cheap fuels added cars in cities without adequate infrastructure, causing traffic gridlock. In 2000, impoverished people were living in vast slums rife with unemployment and crime. If emergy equity can be arranged (see Chapter 9), if populations can be reduced (see Chapter 10), and if the necessity of reversing trends is recognized with the start of global descent, these cities could reorganize with the same goals as those in developed countries. With less structure to change, some of these cities could achieve a prosperous pattern for times of descent sooner than cities in the overdeveloped nations.

SUMMARY

Reintegrating cities with their region of support and influence may help solve severe urban problems while preparing those cities for the decentralization expected in the time of descent. While cities are still growing in many countries, the reorganization to use less fuel is already underway in many developed cities. Decentralization is aided if reduced urban populations can cluster around smaller more dispersed centers. Towns are already developing at the beltway–trunk line junctions, starting with the shopping centers there. For the economy in descent, cities can be kept prosperous by people moving closer to their jobs, by using bikes more, and by reducing the use of cars, parking space, and fuels. A public works program can help make changes in city structure while keeping the poor and unemployed in the economy. We may expect decentralized cities to have less-intensive fuel consumption, less transportation, less strip development and advertising, a smaller percentage of a region's population, a better cycle of materials between the city and its environmental surroundings, a longer rotation of building and renewal, and a more efficient spatial pattern. Greens, including wetlands, ponds, parks, retention basins, and play fields, can be added to depopulated areas.

The inner city is likely to retain its role as a center of information, serving the region through communication technology. The landscape shown in Figure 7.7 is expected to change toward the one in Figure 7.5. Table 14.1 summarizes the suggestions for adapting cities to conditions of descent.

Table 14.1—Policies for Cities in Descent

City and Regional Reorganization

Help orderly decentralization of cities by adapting plans, zoning, and incentives to develop surrounding towns (secondary centers). Operate frequent public transport connecting the central cities with the towns.

Include the central city, its suburban towns, and the daily circulation of people within the same regional government and taxing authority.

Adopt patterns now that will be necessary when rising fuel costs eliminate most private cars. Limit strip development and unnecessary advertising.

Limit the land speculation that inflates prices and causes banking collapse when expected growth does not occur.

Adjust appraisals and property taxes to the money circulation, decreasing rates if there is deflation.

Central City

Keep the centers of information in the central city (libraries, schools, finance, universities, government offices, and computer centers). Use communications to connect the centers to the people of the region (television, the Internet, phone networks).

Remodel and use the buildings that become vacant with decentralization.

Restrict cars from centers of information (high empower density).

Surrounding Towns

Restructure the spatial distribution of housing, businesses, and transportation so that housing and environmental amenities can be near the jobs.

Provide incentives and public initiatives to reorganize housing and transportation around the towns.

Encourage diversity of business in smaller centers to help the one-stop shopping that uses less transportation.

Transfer unemployed and homeless people to public works jobs in redeveloping smaller towns.

Environment

Encourage people to reuse products, to recycle materials from consumers to industry, and to disperse only the dilute, low-transformity wastes to the environment.

Include rural wetlands as part of city operations for recycling storm waters and wastewaters back to nature (compatible with decentralization).

Perforate sidewalks and parking lots to increase percolation of rain so that trees can grow larger. The area of roots receiving water and nutrients needs to be the same as the crown of the mature trees.

continued on next page

Table 14.1—*continued*

Emergy Evaluations

Use emergy evaluation to select land uses with maximum public benefit (different from maximum monetary profit to individuals).

Use empower density and transformity to associate activities of compatible intensity (a means of zoning).

Arrange industries and businesses in appropriate locations according to the city and regional map of transformity and emergy investment ratio. Use centers for concentrated activities and the periphery for more dispersed functions.

15

RESTORING WATERS

At the culmination of a long evolution, the Earth's landscape and its human settlements organized around the hydrological cycle of water. But in the fossil fuel era waters were diverted from their natural processes by development projects. This chapter explains the way the *waterscape* will need to be restored to maximize productivity in times of descent.

Before the era of fossil fuels, human settlements adapted to the higher-transformity zones in the hierarchy of streams (Figures 7.3 and 7.4). For example, the prosperous civilization of ancient Egypt on the Nile River was adapted to use the water pulses, wetlands, and fertility of the coastal delta. It generally made sense for human society to use the natural water pattern without much change. However, in the twentieth century, waters were changed to fit the fuel-based urban economy. Most people assumed—incorrectly—that natural waters were an unused resource and that nothing was lost by diverting waters to power dams and cities.

GREENHOUSE ACCELERATION OF THE WATER CYCLE

While the fossil fuel economy changed water patterns locally, the greenhouse effect—caused by increased carbon dioxide—accentuated the hydrological cycle globally. Evaporation from the ocean accelerates with temperature increase (Figure 15.1). More water vapor and energy goes into the atmosphere, increasing intensity and rainfall in storms, but also making interstorm areas larger, causing longer periods without rain.

However, as fossil fuel use declines, the greenhouse effect will begin to decrease *if reforestation also occurs,* although some years will be required for reabsorption of the gases by the forests, the ocean, and the alkaline desert soils.[1]

Religion and Water in Bali

The link of society and hydrology is so fundamental that it sometimes becomes a part of religion. For example, over many centuries an environmental water system in close symbiosis with people developed in Bali, Indonesia. Using the energy of its elevation, water from high mountains drained downhill, branching into hundreds of small, regulated channels and then into terraced rice fields over much of the island. Rice is a domesticated wetland plant, which can absorb some wastes from human runoff. Faith in a fruitful system of alternating water to different areas became part of the basic beliefs of the culture. Water distribution was managed by religious custom; religious temples marked water control structures.[2]

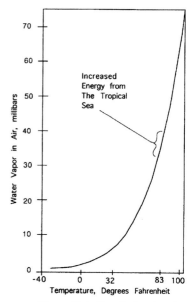

Figure 15.1. Graph showing the increased atmospheric water vapor and its energy generated by increasing the temperature of water.

THE NATURAL WATERSCAPE

The world water cycle uses global emergy from the ocean to organize landscapes. Rain falls from stormy skies, goes into the groundwaters, into lakes, and down the rivers to the ocean, where the sun's energy lifts water vapor back into the clouds again (Figure 7.3). In collaboration with the vegetation, waters carve the lands into drainage basins. After ecosystems make soil, waters move some of it downstream as sediments. As erosion takes weight off the land, more rises up from the deeper earth below to take its place. In other words, part of the work of the geological cycle comes from the sun-driven circulation of water.

Before industrialization, the self-organizing processes of the natural landscape usually did what the human economy needed: conserving water quantity, maintaining water quality, and increasing biotic production, while converting the water flows' energy into useful work on the geological substrates. The ponds, lakes, glaciers, groundwaters, and wetlands stored waters, regulating floods and extending seasons of productivity.

During economic growth, a special-purpose water development could increase empower by combining waters with fossil fuel–based inputs purchased from elsewhere, even though some of nature's work was displaced. It has been common practice to divert waters here and there for short-range purposes and profits without regard to losses of public value. For example, boat navigation and trade has been aided by building levees and channeling rivers, but this change diverted sediments and waters from floodplains with loss of soil fertility and forest productivity.

In the future, when fewer fuel-based inputs are available from the economy, more real wealth may come from allowing the water flows to reorganize a productive pattern. Then human settlements and commerce can be expected to concentrate again where stream flows and tides converge emergy, namely at the mouths of rivers (Figure 7.4), at the foot of mountains, and in coastal estuaries.

Upstream-Downstream Processes

Using the rains, vegetation covers the mountains slopes, weathers the rocks, and in the process makes soil. Torrential streams carve basins, support fisheries, and send high-quality waters and sediments to make valleys and coastal zones fertile. By spreading out the nutrients, sediments, and water in flood plains and coastal distributaries, rivers generate bountiful biotic productivity in wetlands and in estuaries. The rivers increase the productivity of coastal fisheries. Over the years, the geopotential energy of mountains builds valley shapes that concentrate waters and cause floods to surge, doing more geological work and contributing high-quality nutrient materials downstream. The problem for humans is not to disable these pulses but to adapt to them. Instead of preventing floods, human settlements should be arranged so they are not damaged.

Wetlands

Wetlands are lands that are under water part of the year and develop marsh or swamp vegetation (rooted plants that can live in soil without oxygen). Wetlands and beaches filter the waters. Wetland peat has remarkable brown lignin compounds (from the decomposition of trees) that absorb organic toxins and heavy metals.[3] Waters drain into wetlands where the vegetation, peat, and accumulated sediments filter and recondition the waters. The human economy can use these natural filters as low-cost water treatment systems. They operate on solar energy. The world's wetlands, said to be the kidneys of the landscape, are one of the planet's main water quantity and quality regulators.

Some wetlands stay wet all year because superficial groundwaters drain into them. But other wetlands on elevated plateaus conserve waters by collecting rains and transpiring less than those in floodplains. The waters in the elevated wetlands flow out steadily as the headwaters of rivers such as the Suwannee and St. Mary's Rivers, which flow out of the Okefenokee Swamp into Florida. Some of the water in elevated swamps may recharge groundwater reservoirs.

Desert Reservoirs and Underground Storage

In deserts, natural water flows are in canyons and are stored underground. Except during floods, water flows through the sands of streambeds out of sight. Space for water between gravels is about 40 percent and among sand grains 20 percent of total volume. Waters stored there are readily pumped to agriculture or towns. Water evaporates rapidly from open desert reservoirs like the Nile River's Lake Nasser, dammed by Egypt's Aswan High Dam. Open reservoirs in deserts generally waste water. It is better to emulate nature with belowground storage.

Water Law and Conservation

In the western United States, a pattern of law developed in the early days of colonization by Europeans that gave water rights to the first users. This mechanism is analogous to the arrangements in nature that give initial development to the ones that can exploit an unused resource most rapidly. This maximizes empower during early growth. Later, priority shifts to an organization that maximizes efficient use. Law tracks real world needs. As growth–limited by fuel prices and water shortages–comes to an end, replacement of western water law can be expected. Laws will be passed that maximize the whole system of humanity with nature.

REORGANIZATION WITH WATER

During descent, when special uses by the economy decrease, letting the water levels go up and down seasonally can restore lake-margin wetlands. Letting streams rebuild meanders can restore wetlands that were destroyed when those streams were dredged and channeled. We use emergy here, as with all policies, to select alternatives that generate the most emergy.

For example, the Kissimmee River in Florida discharges into Lake Okeechobee. In the original pattern, summer rains in its watershed took three months to pass through floodplain meanders, arriving in the lake as filtered water at the beginning of drier seasons. In the 1960s, the U.S. Army Corps of Engineers channelized and shortened the river, allowing summer rainwater to arrive when the lake level was highest in hurricane season. Water had to be discharged wastefully into the ocean and the Gulf of Mexico. Now a major project is restoring the floodplain, an example of helping reorganize water management to maximize emergy.

Channels, canals, and river diversions were built to help commerce, but required enormous resources to maintain. Our emergy evaluation of the Mississippi River shows that in order to channelize for barge navigation and drain wetlands for soybean farming, many unrecognized benefits of the river's work for the economy were unnecessarily displaced.[4] Some of the levees of the Mississippi River need to be removed to allow the river floods to spread out in larger floodplains again, to enhance the fisheries, to enable the waters to be cleansed by the wetlands and to keep the soils from washing out to sea. With fewer resources available to prevent it, parts of the Mississippi River will revert to their natural channels on their own. For example, the lower river is switching away from the New Orleans channel to the Atchafalaya River channel. When fuels are expensive, the economy will have to fit the natural delta pattern. The sooner this is allowed, the sooner New Orleans will be protected from the catastrophe of flooding from a major hurricane.

In times of rich resources when buying power of salaries and pensions was high, tourist and retirement cities developed. These new cities had large demands for water, much of it for luxuries such as swimming pools and ornamental vegetation. As fuels for the cities become costly, waters will go back to agriculture and to the green life-support cover of the Earth. As the cities lose their buying power for luxuries, their people will depend on the rural basis of their region once again.

Waste Dispersal and Wetland Recycling

Wastes not reusable by the economy can be returned to the natural processes in form and concentration that can be used beneficially. Return of waste materials to nature's landscape usually requires diluting them so their concentration in any one area of an ecosystem is not beyond the ability of the ecosystem to consume and denature the waste substances. The famous phrase "the solution to pollution is dilution" is partly right so long as it does not mean putting pollutants into streams that converge downstream and reconcentrate pollutants. Dilution has to be accompanied by dispersal through nature's means of recycling, such as through wetlands or beaches. By restoring wetlands, river systems can maintain their water quality, reducing human illness and the costs and taxes for water treatment.

Some regulations require complete treatment of municipal and industrial waste. Primary treatment removes solids; secondary treatment removes organic matter; and tertiary treatment removes nutrients. However, the chemical process for tertiary treatment of sewage effluent is very expensive and does not use the nutrients for useful purposes. An alternative is to use wetlands to absorb the nutrients. Experiments using freshwater and saltwater wetlands have been very successful in many countries. Pumping wastes into wetlands eliminates

*FRANK & ERNEST by Bob Thaves © Courtesy of United Feature Syndicate.
Reprinted by permission.*

pollution. Artificial wetlands are easily constructed so as not to disturb
natural areas. Wetlands can be arranged between most wastewater
runoffs—from our cities, industries, and agriculture—and natural waters—lakes, rivers, and estuaries.[5]

For example, in Florida, treated sewage effluent has been piped
into elevated swamps of cypress and gum trees. The trees increased
their growth using the nutrients in the water, and the water, which
eventually percolated into the groundwater, was essentially free from
nutrients, heavy metals, and disease bacteria and viruses.[6]

Nutrients and residual microbes in treated sewage are detrimental
going directly into streams or estuaries. However, our studies in North
Carolina and Florida showed that when wastewaters are released
through marshes and mangroves, the excesses are filtered, and organic food matter is released to estuaries gradually in a more natural
form. Fisheries are stimulated, and treatment costs reduced.[7]

Aquaculture

Both freshwater and saltwater ponds are important for raising fish
for harvest. Some traditional, low-intensity aquaculture is done with
very little effort by humans other than to harvest the fish crop, but
other systems are even more intensive than industrial agriculture. Scarcity and rising prices of fishery products make intensive aquaculture
competitive, sometimes catering to consumers affluent enough to pay
for special delicacies. For example, thousands of saltwater ponds in
coastal Ecuador supply shrimp to developed countries. Costs are relatively high for a rural area because so many purchased inputs are
used, and the local workers cannot afford to buy the shrimp being
raised in their own area.[8]

Salmon are grown in floating pens in estuaries of British Columbia and Norway. They are fed with shredded fish and crowded like
cattle in feedlots, except seawaters receive the wastes. The young salmon
come from hatcheries that get their eggs and sperm by diverting

mature adults away from their normal streams. With too many pens, waters are polluted and estuarine values are lost.

Like agriculture, a declining investment ratio will shift aquaculture toward less-intensive operations. It is likely that aquaculture will increase but gradually shift away from luxury products, growing instead species lower in the food chain with greater efficiency and less cost. For example, aquacultured catfish and tilapia, both low in the food chain, are already in the grocery stores. Some ponds can be managed to produce fish food by their own photosynthesis or to use plant materials cut from surrounding lands. Costly high-transformity protein fish food will be used less. Judging by its earlier wide use in densely populated countries, low-intensity aquaculture is an efficient way to obtain proteins while absorbing nutrients and by-products of agriculture.

DAMS AND POWER

Dams, large and small, were built to store water for many purposes: for cattle, crop irrigation, aquaculture, flood catchments, barge transportation, town water supply, sediment and nutrient traps, boating recreation, and aesthetic vistas for housing. Whether the dam is a net benefit requires an emergy evaluation.

When purchasable resources were cheap, development of a dam on a river could attract economic developments that combined the emergy of water use and that of imported fuels. The development could generate a greater emergy contribution to the economy than the original river pattern. However, as the imported resources become expensive and less available, the river's original role will generate more value.

In the past when a river was dammed, channelized, or diverted, the economic contribution of the development was calculated, but the loss of the indirect contributions of the original system to the public economy was not included. Evaluations used market values of fish and water as raw commodities. However, a river's free indirect contribution to the gross economic product in food and forestry products, in renewing soils, filtering wastes, reconditioning drinking waters, doing geological work, and maintaining coastal economies, is much greater than its market values. In most instances, the emdollar value of waters is several times greater than its market values.[9] Thus the cost-benefit analyses incorrectly favored the projects.

Hydroelectric Power

After damming, electric power is produced from the elevated water plunging through turbines, but there are losses of older values such as food from migratory salmon, shad, and sturgeons, passage of sediments

to valleys, and floods that stimulate marshes. Whereas many one-pur-
pose developments can be dismantled during descent, power dams
are an exception. Hydroelectric power is essential to maintain information networks for national and global unity when fuels are less available (see Chapters 13 and 17). In 2000, many projects in Oregon and
Washington sought to keep salmon running without removing dams.
Juvenile fish are killed in the turbines as they go downstream, and not
enough returning salmon have passed upstream in fish ladders. (A
fish ladder is a narrow branch of the stream that goes around the dam.)
Many dams have no ladders, and some ladders are too steep and
without resting pools.

RESTORATION OF COASTAL FUNCTIONS

Coastal areas have their economies enriched by the emergy inputs
from river flows, organic matter, sediments, energy from freshwater-
saltwater mixing, work of tides, winds, sea-level changes, waves, and
beaches. The sea works for coastal economies without being paid.
During the fuel era, many of these resources were reduced to fit fuel-
based investments, such as estuarine mining (oil, oyster shell, dredged
sand), barge traffic, and ocean-going trade. When fuels become less
available, values can be increased by helping the system of waters and
tides to self-organize together.

Often, economic developments divert the original values of marine systems from the public economy. For example, to make bridges
cheaper, causeways were built on fill, but that fill blocked tidal circulation and reduced the ability of the estuary's biological system to
receive and clean up wastes. Consequently, fishery values were reduced, tourism decreased, and higher taxes were required for waste
treatment. During descent, reorganizing human structures to encourage more of the natural marine resources can help the human
economy.

Estuarine Nurseries

Many crabs, shrimp, and marine fish important to commercial
and sports fishing breed far from shore, after which their eggs and
larvae are brought by the tide into the estuaries, where they do their
fastest growing. Thus, estuaries are called nurseries and are essential
to bountiful fisheries. Conservation of estuaries with their marshes or
mangrove swamps sustains fishery production.

Rivers bring to estuaries freshwater (with its chemical potential
energy) and fertilizer nutrients like phosphorus and nitrogen, which
make the marshes grow better and increase the productivity of the
fish-generating coastal ecosystems. Rivers also bring organic matter from
lands and by-products of agriculture or industry. Estuaries make major

chemical contributions to the human economy by metabolizing and denaturing wastes as they produce fish.

Oyster Reefs

Oyster production is dependent on the surge of river flow because oysters require an alternation of low- and high-salinity water that eliminates predators, competitors, and diseases. Oysters build reefs of shell on which they can readily attach to filter food from circulating water. They are killed by reduced circulation, removal of the surge of freshwater, dredging up of their shell reefs, or the introduction of so much nutrient runoff that indigestible eutrophic algae prevail. Where sewage is released, oysters become unsafe for eating and their reefs are declared off-limits. Few places are still safe for eating raw oysters, because, being filter feeders, they concentrate microorganisms, including those causing hepatitis. Oysters near industrial waste can also concentrate heavy metals or radioactivity. We can help renew oyster use by releasing wastes through wetlands before they go into estuaries and by restoring water circulation.

Sediments and Dredging

Many rivers carry sediments that are eventually deposited in the coastal zone. Freshwater wetlands along the river and saltwater wetlands along the margins of the estuary receive the sediments and are nourished by them. However, too much sediment makes an estuary turbid, light does not penetrate, and productivity is diminished.

Dredging for economic developments often has damaged the public economy unnecessarily by blocking circulation, upsetting bottom organisms, and keeping bays turbid. Unless special efforts are made to divert the dredged sediment, dredging of ship channels often leaves spoil islands that block water circulation. Fine sediments are dispersed in the bay and subsequently stirred up with each storm, blocking light and biological productivity. Deep channels dredged in shallow estuaries have pockets of oxygenless water with toxic hydrogen sulfide that is stirred up by passing ships. During descent, transport with smaller ships and barges can fit navigable waters with less dredging.

Fisheries

The stocks of world fisheries have been so reduced by overfishing and pollution that many have collapsed. Capital investments put in too many large, efficient harvesting ships in excess of the renewable carrying capacities of the fish. This made the stocks too small to release enough offspring to make optimum fish yields. As the fish became scarce, their prices rose, which caused even more overfishing. In many areas, the desired fishery system was displaced by less-desirable

"One day, with the money made by industry we'll be able to clean up the environment"

From EARTHTOONS by Stan Eales. Copyright © 1991 by Stan Eales. By permission of Warner Books Inc.

"weed species." Sardines in California and the sea-bottom fishes off New England were overfished and have not returned.

For the time of descent, the lost fishery resources need to be restored by reducing harvesting and pollution. With rising costs of fuels and boats, fishermen will not go so far at sea, operate as large boats, or pull as big nets. Fishery stocks will return to a more productive level. The catch per effort can be higher again.

Although some marine waters such as saltwater grass flats in shallow fertile estuaries are as biologically productive as land crops, most of the sea is not as productive as the land. Many claims in earlier decades about the capability of the sea to feed more people were exaggerated. In spite of much more fishing effort, world fish harvests are not increasing.

Beaches and Jetties

Wave energy and currents that carry sand maintain beach systems. Beaches are zones where the wave energy concentrated from winds over the ocean is received and used. The breaking waves organize the sand into beach form, filter the waters through the sands while cleansing the water, and move sand up and down the beach. To maintain a beach, the sand that is lost out to sea must be replaced. Normally, wave-driven currents move sand from the mouth of rivers where rivers deposit sand or from sand deposits formed by coral reefs.

As long as they get their waves and sand, beaches are tough. Beaches develop dunes that protect coasts from hurricane impacts, contribute to the life-support role of waters, support wildlife, entice people for recreation, and attract economic activity.

During our times of affluence and heavy tourism, short-term changes were made in beaches that were not sustainable even when fuels were cheap. Buildings too close to the surf were destroyed by storms. Jetties to catch sand did the opposite, cutting off the necessary flows of sand up and down the beaches, causing beach erosion. For example, large jetties in the mouth of Jacksonville's St. Johns River help keep the navigation channel open, but cut off the flows of sand to beaches to the south. Beaches, housing, and roads were subsequently lost in storms.

Coastal Groundwaters

In many places the sea is encroaching on the land, undermining hotels and houses. Because of the publicity about global warming, many people think this is caused by the rise of sea level, but in fact the sea has risen only a few inches in the last hundred years. Instead, many lands have sunk because local groundwater has been pumped out. In famous examples in Taiwan and Venice, Italy, uses of groundwater near the sea caused the land to compress and sink. Groundwater use near the sea also draws saltwater up from below, ruining the groundwater for many human uses. Pumping groundwaters near the coast might be restricted to drinking water purposes.

Emergy evaluation shows waves and sands contributing much to the wealth of coastal economies. People take it for granted until they have to pay money for the beach. In south Florida, millions of dollars are spent to pump in sands to traditional beachfront areas, with only a temporary effect. If construction is set back from the beach and jetties are removed, more dune sand will be available to the beach processes.

SUMMARY

For prosperous descent, the economy as it self-designs can achieve more by fitting into the global hydrologic cycle. When there is less fuel to divert waters into special-purpose profit making, more empower is generated by adapting society to the multiple values of rivers, estuaries, and beaches. Human settlements and the ecosystems around the rivers can reorganize so that the economy better fits the water system. Nature's treatments, properly used, work for the economy so that tax-driven technologies don't have to fulfill the same functions. Use floodplains to maintain water quality, restore estuarine circulation, remove jetties, stop pumping coastal groundwaters, operate lower-intensity aquaculture ponds, restore reefs, and set levees back from the shore. Coastal fisheries can recover when fishing pressure decreases and some of the nutritive river discharges are restored. The energies of mountain waters will still be needed to produce hydroelectric power for society's information systems.

16

REFRESHING
THE LANDSCAPE

As cities decentralize and use less fuel, productivity will depend on reorganizing the symbiosis between the economy and the land. Agriculture, forestry, and mining will become less energy intensive, and many people will return to jobs in rural areas. In this chapter we suggest policies for refreshing the biotic cover, productivity, diversity, and human resettlement of the landscape.

As explained in Chapters 8 and 13, the landscape cannot support the present economy unless populations are reduced. Otherwise, the landscape is stripped barren and society collapses. For example, in 1998 hurricanes flooded the overpopulated and deforested nations of Haiti and Honduras, devastating their soils, settlements, and economies.

Because society needs fuels, food, clothing fiber, housing, and other critical materials, a varied mix of land uses is required, each operating with *pulsing* cycles. The landscape mosaic needs areas for *ecosystems, agriculture, forestry,* and *mining.*

ADAPTING TO PULSES

In Chapter 5 we explained the way all systems pulse after building up storage in a pre-pulse period. Since the environment has systems on many scales, the pulsing occurs on short- and long-time scales. Adapting human economy to the landscape means adapting to the pulses on each scale. There is an optimum time of rotation that, in the long run, maximizes farm and forest yield. Farmers alternate pastures between periods of grass growth without animals and short periods of consumption when cattle are allowed to graze. In forestry, humans manage the period of net growth of the trees for some years and then cut and clear prior to replanting. To be sustainable in the long run, use patterns have to be cyclic.

Where environmental pulsing is frequent and therefore with little time to accumulate energy for each pulse, society can adapt by making structures strong enough to sustain pulses. For example, ordinary houses are built to resist ordinary winds and insulated enough to even out sudden temperature oscillations.

Where pulses are large and frequent, humans can adapt by working within natural cycles. For example, farming of rich floodplain soils has to fit between soil-renewing floods.

An observer seeing a forest consumed by fire, storm, or insect epidemic thinks of it as a catastrophe. Yet in the larger scale of things this is regular renewal and the normal expectation. The large-scale but less-frequent pulses of the environment—floods, hurricanes, earthquakes, or volcanic eruptions, for example—are often strong because of the long accumulations of energy between each episode. Human uses have to fit the strong environmental rhythms also.

Where environmental pulses are infrequent and catastrophic, humans can adapt by not building in high-risk areas or by building minimal, low-cost, easily replacable structures. People should not occupy areas where the emergy required for construction and replacement is greater than the emergy produced by using the buildings. Especially in a time of descent, fewer resources may be available for repair and replacement. Areas of high-energy destruction can be mapped and incentives provided to leave such areas to the adapted ecosystems. The economy is unnecessarily drained when insurance policies are provided for building in disaster-prone areas (such as floodplains and ocean beaches).

Disaster Relief

As populations and affluence increased in the 1990s, more and more developments were made in high-risk areas such as floodplains, earthquake fault zones, and beaches. Economic losses from floods, hurricanes, and earthquakes increased. In the United States the Federal Emergency Relief Agency (FEMA) provided funds to help people rebuild after each disaster. For example, great floods of the Mississippi River in 1993 devastated towns that had been built in the floodplains without putting structures on stilts. However, policies began to change. Instead of being paid to rebuild on the floodplain, people were paid to move out. The luxury of the public's subsidizing a poor fit of society to land was questioned. In times of descent, the public repair of ill-adapted structures is likely to disappear.

Impact of Global Climate Change

The greenhouse warming of tropical seas enormously increased the water vapor energy going into the atmospheric system. This is because evaporation accelerates with temperature increase (Figure 15.1).

HI & LOIS by Brian and Greg Walker. Reprinted with permission of King Features Syndicate.

The global effect so far has been an increase in both the size of storms and the length of the dry periods between the storms. Floods are greater *and* droughts are longer. The intensity of east-west climatic oscillations known to the public as El Niño and La Niña has increased. The impact of disasters and agricultural failures increased because of the global deforestation that removed much of the natural protection of watersheds and wetlands.

If reforestation occurs, however, the decline in fossil fuel use will lead to a decrease in atmospheric carbon dioxide, one of the gases causing the greenhouse effect. See Chapter 15.

RESTORING ECOSYSTEMS

Ecosystems are the reservoir of genetic variety necessary to maintain soils, cleanse waters, and purify air, essential for human life support. Ecosystem storages and structures necessary for its productive processes– including trees, plants, soils, seeds, birds, microbes, and biodiversity– are called *natural capital.* But the natural capital of ecosystems has been decimated in many areas. Restoration of ecosystem functions requires that uses be limited until they can recover their natural capital, taking from ten to 100 years. In some areas ecosystems will have to be restarted.

Rebuilding Natural Capital

In agrarian times, emergy of the rural environment was the main basis for human society. Even political power was rooted in land ownership and the rural culture. Then, after 1850, the intense industrial-technological, fuel-based civilization of the cities began to develop. New uses loaded the environment-economy interface (Figure 6.6).

Environmental resources–land, water, air, minerals–were diverted for use with the fossil fuels and the cheap goods and services from the new fuel-based cities. The emergy investment ratio increased. Not only was environmental production diverted to the human economy, but much of the natural capital was harvested also.

Forests and prairies were cleared. Intensive agriculture used up soil through erosion, compaction, oxidation of organic matter, and loss of the soil microorganisms. As much real wealth in soil reserves (measured as emergy) is depleted each year, even now, than is used up from coal reserves.[1] To make the environment productive again, natural capital that was lost in the growth era needs to be restored. This means putting lands into a rotation that includes a period when ecosystems restore soils. (See p. 242.)

Wildlife and Hunting

The larger animal species are part of the natural capital, but preserving them is difficult. Hunting has become an outdated relic of frontier times when populations were sparsely distributed. In many areas there are more people than wildlife. There is not enough wildlife for people to supplement their food supplies with hunting. There are other means of enjoying wild areas. The majority have rights to visit the few remaining wild areas with their children and see reasonably approachable wildlife.

Hunting is not a safe practice where people are using the environment for other recreation. There are too many hunting accidents. Many people still hold to the old ways that use hunting to teach the use of firearms, but use of firearms is better taught on rifle ranges. For the wild areas, native carnivores are better than hunters at controlling the oscillations of deer. Excess deer in housing areas can be driven into enclosures for removal by the venison industry.

Larger animals that have large territories survive better if the remaining wild areas are interconnected by strips of natural vegetation. Suitable strips are already available if allowed to grow vegetation cover: the riparian borders of waterways, the margins of highways, the lands under power lines, and those over pipelines. Highways need underpasses to reduce roadkill and allow the genes of mobile populations to exchange. In 2000, wildlife underpasses were added to Florida highways crossing the Everglades and the wetlands of Paynes Prairie.

Exotic Species

In the past century, much of the Earth's land surface, soils, and aquatic habitats were so changed by human development that the previously adapted successional sequences of species were no longer well adapted to restore productive ecosystems. However, many species of plants and animals from the Earth's biodiversity have been established in other areas aided by humans' crisscrossing the planet in ships, trains, cars, and planes. Consequently, many disturbed areas were taken over by highly productive exotics that were adapted to the new conditions. Some argued that these invasions are harmful and are displacing native species. Introducing exotic carnivores did cause extinction of some native animals. For example, the introduction of carnivores into New Zealand caused the extinction of flightless birds.

However, usually the exotic plants and animals make an area productive for which there were no natives adapted. Natural self-organization can accelerate restoration by using exotics. For example, in Florida, draining water tables produced dry lands for which exotic species from dry-season climates were better adapted than the original wetland species. Restoring water regimes often restores original ecosystems. Exotic plants capable of using and binding the excess nutrients perform a useful service in some areas made eutrophic by fertile runoffs.

Exotic trees in Puerto Rico rapidly reforested most of the island after agricultural lands were abandoned. Then native species, using the soil improved by the exotics, began to develop underneath. Exotics often serve the long-run needs by achieving the maximum possible productivity during stages of restoration. In general, adding exotics increases the diversity of species available to ecosystem development. However, on isolated islands like Hawaii, exotics can cause the extinction of genetically limited and less-efficient native species.

TURNAROUND OF AGRICULTURE

Until now, agriculture has made progress by using more purchased inputs. Research and technology turned agriculture into an industrial process using machinery, fuels, fertilizers, pesticide chemicals, and intensive business methods. Yields per acre were increased many times. A few farmers operating larger farms supplied more food and fiber than many more farmers did earlier operating small farms. Cheap fuels and fertilizers made possible higher productivity and larger-scale operations. Displaced farm workers went to the cities where they often produced the equipment and chemicals used in the intensive farming.

But during descent, with rising prices making fuels and fuel-based inputs less available, crop yields per acre will decrease, and more land area will have to go back into agriculture in order to provide enough food. Purchased inputs will have to be reduced and production per acre has to decrease accordingly. As less-intensive agriculture becomes more economical once again, farmers will be forced by costs to use less equipment, chemicals, and irrigation to keep their market prices competitive. When purchased inputs decrease, a higher proportion of the work of growing the crops falls on the environment. Organic farming is already increasing.

More planting, cultivating, and harvesting will have to be done by hand, requiring more labor and animals for less-intensive farming. Farm employment will increase, and costs of food and fiber will take a higher percent of consumers' income. A program of migrant labor could be arranged, not from other countries but from the city unemployed. Many idealistic young people in the 1960s prematurely attempted lower-intensity farming. Soon the rising food prices associated with less fuel and soil availability will make it more economical.

During times of growth there are unfilled demands and rising prices, and the farmers that borrow move into an open opportunity quickly. But in times of no growth, farms with large loans will not compete with those farms without the loans. The high interests they pay will raise their costs, and make their operations less competitive.

The agricultural turnaround will show up as a gradual decrease in the investment ratio (the ratio of purchased emergy to the local free environmental emergy—Figure 6.6). By tracking this ratio as it declines, people can plan for the appropriate intensity of agriculture. An intermediate stage would use smaller farm equipment and fewer purchased chemicals. A century or more could pass before the investment ratio of agriculture decreases from 7 to 1 or more in 1999 to values of 2 to 1 or less for agriculture based only on the land.

REINFORCING THE LAND

From Chapter 2 we are reminded that a sustainable economic use requires feedback reinforcement to keep the environmental system, especially the soils, healthy. Figure 6.5 does not have reinforcement from farmers or the economy going back into land processes, but Figure 6.7 does. Under stringent free market competition, those who omit care of the land win out in competition in the short run, but their farms lose fertility and become uneconomical in the long run. For example, putting land into fallow for restoration of soils is costly because the lands still have taxes and costs of protection and management. Incentives such as tax exemption for restoration and upgrading land can yield more emergy to society than is in the tax incentive money.

Farming draws part of its environmental emergy basis and hence economic activity from the soils, waters, and the organization of the landscape. But the money that farmers receive for their crop in competitive markets is only for the costs of human services that supply goods, services, and labor. Money from sales does not cover the work of the environment, yet environmental processes are needed to maintain the natural assets of the farm (Figure 6.7). Otherwise the farms deteriorate and lose their productivity.

Subsidies—the feedback from consumer society to reinforce farms (Figure 6.7)—are necessary to sustain agriculture. Some kind of subsidy is found in many countries. The idea is that money allows the farmer to manage the land to sustain yields. For example, the farmer may be paid to terrace land to reduce soil erosion. Even in 2000, keeping an agricultural landscape was considered part of the environmental balance with urban development. Maintaining some agriculture preserves the locally adapted agricultural varieties that may be needed later. Sometimes a subsidy to farming has been justified as a national security policy to make sure the country can sustain itself if sources of imported foods are cut off by war. For example, agriculture was supported in Switzerland, Sweden, Japan, and Bermuda. National policy in Taiwan prohibited urban development of farmland around urban Taipei.

In 2000, world food supplies were less secure because of the loss of agricultural lands due to erosion, desertification, and encroachment by cities and housing developments. Globally, stocks of food in reserve were less. Rising costs of fuels will reduce the international transport of bulky foods. As the decreasing availability of fuels causes yields per acre to decrease, each country must again produce more of its own food. Agrarian economies are practical because crops, aquaculture yields, and forest products usually contribute net emergy.

Fertility

Figure 7.1 shows the necessary return of materials such as fertilizers to the landscape. Fertilizers are used to replace soil nutrients that are depleted—when land is farmed. Nitrogen fertilizer is generally produced from the atmospheric nitrogen using fuel energy. Potassium fertilizer is obtained from salt deposits and the evaporation of brines. Phosphorous fertilizer is mined from rocks in Florida, Idaho, and Morocco. As time passes, the largest deposits are being depleted. Costs increase for the fuels and electric power operating the giant draglines that dig up the phosphate. As these minerals become less available and more expensive, fertilizer prices will rise. When farmers can no longer afford to buy as much fertilizer, it will pay to use more natural ways to conserve the nutrients in the soil.

Nitrogen is put back into soils by growing plants—soybeans and clover, for example—whose root nodules can change the nitrogen gas in the air to usable nitrogen in the soil. Small amounts of phosphorus and potassium accumulate when fields are left fallow. All the fertilizer nutrients—nitrogen, phosphorus, potassium, and the many others needed in small amounts—are recycled into the soil as plants in the fields die and are decomposed by microorganisms. More agricultural and urban nutrient wastes, manure, and compost can be recycled into the soil.

Land Rotation

Especially when the inputs to industrial agriculture decrease, the land rotation system is the only way to maintain fertility. Productive systems generally operate best in pulses (see Chapter 5). Perpetual fertility is maintained by rotating land from agricultural use to a fallow period when soil nutrients are restored by complex natural vegetation. Agriculture uses up soil, but forest succession rebuilds its fertility. Rebuilding soil is one of the best uses of solar energy.

If strips of high-diversity ecosystems exist nearby, land left fallow is automatically reseeded with the help of the birds and other animals. Long-term sustainability requires that about 10 percent of the land be kept in patches or strips of mature ecosystems. These areas can also serve as parks. Without these sources of life, regrowth and recovery of farmed land is delayed. There is clear economic advantage from prompt regrowth of complex vegetation in the fallow part of the cycle.

Many areas in the eastern United States and western former Soviet Union regrew forests and soils rapidly after agriculture was abandoned because patches of high-diversity forest had been saved throughout these regions. Few plots were more than a fraction of a mile from a seeding source. Similarly, the old pattern of slash-and-burn in tropical forests worked because the animals and seed sources were there, the density of farming was small, and the time between cutting long.

In contrast, large areas of New Zealand were clear-cut and natural seedlings were eliminated by sheep. When these lands were abandoned, their recovery was stunted in a vegetation scrub (called gorse) for many years because plots with the native seed trees were too far away.

The government of Brazil tried to maintain its biodiversity in some areas by saving half of its land in complex diverse forest around its main economic development projects. In the Atlantic rain forest in Brazil's Bahia area, forest in moderately diverse second growth is in long rotation with agricultural tree crops.

Where lands have new fertility continuously added (such as floodplains), lands can be farmed more frequently and with shorter rotation. Lands that have to accumulate nutrients from rains, because their rocks do not contribute much, require a long fallow period. Maps that show the appropriate rotation time for each land type can help farmers.

Agricultural Varieties

In the twentieth century, higher-yield crop plants and farm animals were developed, but these varieties required more care and cost.

What was done successfully was not an increase in conversion of solar energy, but a diversion of the photosynthesis into usable parts—fruits, grains, and fibers. Many plant functions, such as resistance to disease, drought, temperature extremes, and pests, were bred out to increase yield. The high-yielding varieties required more care and more inputs, which were supplied directly and indirectly from cheap fossil fuel. Many agricultural leaders have never really realized nor explained to the public that the organisms were giving up something in exchange for increasing yields.

Miracle rice is a famous example of a plant with increased yield and reduced self-sufficiency. The higher yields of miracle rice required high levels of management, a program of breeding rice varieties to keep up with pests, and costly inputs of fertilizer, pesticides, and irrigation water. The high-yield rice agro-ecosystems were a way to convert cheap oil into food.

Changing the genetics of microbes with biotechnology methods has successfully generated unique but high-priced pharmaceuticals. To succeed, genetically engineered crops, plantation trees, and farm animals have to operate for at least a season in the outdoor environment in competition with wild forms that use more of their energies for survival. More efforts are required to protect the specialized varieties.

Biotechnological changes in crop genes need to be consistent with the whole system of the environment and society. For example, adding genes to make corn toxic to insects may interfere with consumer biochemistry and normal decomposition. The natural pesticides found in tropical root crops are readily washed out and decomposed.

In the rush to develop the high-intensity producers with their excessive costs, many of the lower-yielding varieties—those that require less management, pesticide, fertilizer, and irrigation—have been discarded. Priorities are needed for agricultural experiment stations to identify intermediate-intensity varieties and supply recommendations for their use in intermediate-intensity agriculture. It is important to retrieve these lower-cost, less fuel-consuming varieties before they are lost. Conversely, the high-yielding agricultural varieties that require high energy inputs also need to be kept alive in genetic banks preserving information for the future (see Chapter 17).

FOREST MANAGEMENT

Several kinds of forest management areas will be important for a landscape in the time of descent: (a) old-growth, complex forests, (b) forest plantations cut for yield, (c) crops cultivated under the canopy of overstory trees, and (d) orchards. Reforestation is already recognized as a global priority.

Complex Forests

Complex forests restore soil, preserve biodiversity and wildlife, provide aesthetic experiences, prevent soil erosion, and slowly grow high-quality woods. After many years of development and self-organization, the emergy stored in these "old-growth" forests is very large.

Under the pressure of economic policy to make short-range profit, most of the world's old-growth standing forests are being stripped without provision for their replacement. Incredible as it seems, bankers have advised cutting all the forests and putting the money in trust funds with interest to take care of the future. The fallacy is that the value of money depends on continued production of real wealth by the environment. The interest they were counting on requires growth in resource use by the economy not expected during descent. Destroying environmental systems that are the sources of real value reduces the buying power of money and the interest. During descent, the new wealth from forest growth can be greater than that from interest. (See *discount rate,* p. 199.)

Although natural, old-growth forests have more gross photosynthesis than young forests, and they grow more slowly because they put most of their energies into quality structure, diversity, resistance to insects and weather, recycling, and reproduction. The net yield of wood per acre per year is less than that from plantations. However, old-growth natural forests with many years of wood storage require little input from the economy, except protection and cutting services, and thus yield ten to 100 times more to the economy than they draw from it. The longer the environmental production is allowed, the greater the net emergy yield. We sometimes say: *time is emergy.*

The valuable woods of complex natural forests have higher density than plantation wood. With many special qualities for furniture, construction, strength, and long life, natural woods are the most in demand. They are still coming from the old-growth forests of the world, which are rapidly disappearing. Almost no effort is being made to replace these forests because of the fifty to 100—or more—years required to grow the higher-quality products. Valuable woods are being mined, not renewed. When fuels that make plastics are gone, we will have to use the higher-quality woods more, even though we will have to wait for their slow growth. It takes more time and energy to grow quality. Society will have to adapt to the forest cycles by reducing demand.

The area of land and network of streams that catches rain and carries it downstream is called a *watershed.* Watersheds used for water supplies need to be kept in complex forests so that the rainwater running off is high quality and not eroding the soils faster than they form. When toxic substances are in the rain because of air pollution, complex forests often absorb them. Bare-land management on mountain watersheds threatens public values. Complex forests have many small gaps (tree falls) and fewer larger ones (land slides). Tree harvest methods that keep the heterogeneous gap patterns are more natural than the clear-cutting of large tracts.

Bare-land farming, clear-cutting, and paving of steep slopes overload streams, eliminate vegetation needed for water management, erode soils, destroy fish habitats, cause floods by shooting water downstream too fast, and prematurely fill dams with sediment. All these practices diminish the carrying capacity for the landscape to support the economy.

Forest Plantations

Forest plantations will have to supply wood for construction, paper, house heat, fruits, nuts, and spices. Like agriculture but on a longer time scale, forest plantations' sustainability depends on a rotation of plots between growth, harvest, and replanting. Wood grown as an agricultural product requires about twenty-five years from planting to cutting in warm climates' long growing seasons but about ninety years in colder climates. In contrast to natural forests, plantations' energies are channeled into one or two products instead of into diversity. Some plantations grow more than one kind of tree for yield at the same time, creating a more diverse harvest.

The wood from rapid-growth forest plantations is low density. Its energy content is less than in mature natural forests, and many of the properties of strength and long-term service required of a building material are missing. However, fibers of pines and other species are good for paper production. After the first growth cycle, fertilizer is often required. Weeding is sometimes needed while seedlings are small. How long tree plantations can be maintained without rotating through a period of regeneration by a natural forest is much argued.[2] It depends on the climate and the type of soil.

Stronger, more versatile, and longer-lasting building materials are made from these lightwoods by various combinations with plastics and synthetics from petrochemicals. Since these ultimately depend on oil, they will be less available after fuels become expensive. Plantations require inputs of fuel, fertilizer, equipment, and human services. The emergy yield is only two or three times the inputs used in the process.

The more a single species is used as a monoculture, the more hazard there is that an insect pest or plant blight will develop into a destructive epidemic. Diversity of vegetation, birds, and insects eliminates epidemics. Alternating plots of different kinds such as agriculture, forests, parks, and human settlements also helps prevent epidemics. Many plantation forests have a single species overstory like pine but a highly diverse self-seeding understory of plants and animals, which gives the forests more protection from insect outbreaks.

When fuels were cheap, plantations contributed to an economy most by using fuel-based inputs to increase yields, a way to convert rich fuels into more wood products. However, with declining resources, letting a more complex forest generate wood with less costly inputs becomes an economic advantage. In that case, having adjacent patches of old-growth complexity provides prompt seeding after harvest, making cycles of growth and harvest shorter and more economical.

Many forest practices driven by short-term profit motives are inconsistent with continued productivity. Tree harvest techniques that pull up the roots hurt the soils, land texture, and future productivity. Clear-cutting on steep mountains loses soils to erosion. Using bulldozers to shove brush into rows after cutting shoves fertility into the rows that grow weedy competitors rather than the intended trees. Burning brush piles removes the nitrogen and organic matter that the soil organisms need to maintain soil quality. Removing organic litter (biomass) for fuel use does not maintain soil fertility.

Crops under Overstory Trees

Especially in the humid tropics, special food products like coffee and cacao (the raw ingredient for chocolate) were grown as bushes or small trees under overstory trees that provided appropriate conditions. The overstory trees maintained soil and water quality, a stable microclimate, nutrient cycles, and biodiversity control of pests. Some branches were trimmed each year to admit sunlight. With industrial agricultural methods, these same crops are cultured as a monoculture in full sunlight without the overstory trees. Yield is increased, but everything the overstory did before has to be supplied at much increased cost (fertilizer, pesticides, labor). With fewer purchased inputs available during descent, the older overstory system will probably be used more again.

Orchards and Dooryard Gardens

In the late twentieth century, extensive single-species orchards and vineyards were the source of nuts, oranges, apples, bananas, grapes, and many other fruits sold in world markets. Operating these orchard monocultures is energy intensive, requiring many inputs of fertilizer, machinery, chemicals, and services. In traditional ways of life in earlier times, however, many kinds of fruit trees were included around human settlements, stimulated by recycling of household wastes. In Indonesia, Panama, and pioneer settlements of the United States, these were part of the dooryard gardens. In times of descent we may expect the re-emergence of diverse fruit tree cultivation within human settlements.

Reforestation

Restoration of forests is now a critical global problem. In many places no seed trees are left. Changing climates and the carbon dioxide accumulating in the atmosphere are caused not only by the burning of fossil fuels, but also by the presence of fewer forests to use carbon dioxide through photosynthesis. Forest destruction is so serious that many overpopulated countries do not even produce enough wood for their cook stoves. There are several ways to restore forests:

- Require those cutting forests to leave enough seed trees and diversity patches to reseed automatically.
- Set aside lands at the edge of existing forests so that the forest organisms can spread by means of wind, birds, bats, and other animals.
- Restore strips of high-diversity forest so that birds, bats, and ground mammals can carry seeds and microbes to bare lands nearby.
- Allow old plantations to become reseeded with the natural biota to change them into complex forests.
- Allow weedy exotic trees to cover the bare lands first, providing conditions of shade and soil for the regular forests to develop underneath.
- Plant seedlings obtained from tree nurseries.

MINING AND MATERIAL RECYCLING

Most of the nonrenewable resources that supported our century of growth were mined from the reserves stored by slow geologic processes over long periods of time and gradually brought to the surface at about an inch per thousand years. The very concentrated rich deposits close to the surface were mined first and with little effort.

But now the net contribution by mining is decreasing. The lower the concentration of a mineral and the deeper the deposit, the more earth has to be disturbed and put aside to extract it.[3] Costs of mining are greater, and more land is taken out of production for longer periods. The required fuels and electricity are becoming more expensive. As the rich material resources available close to the Earth's surface are used up, the relative roles of mining and reuse change. More and more things are reused, recycled in industry, or returned to the environment to be processed in a way that is productive to the renewable economy. Reuse and recycling replace mining (see Chapter 10).

Net Emergy Limits

During descent when fuels are expensive and the remaining deposits are poor, deep in the ground, or far away, costs of mining will be high, and people will use the products less. Most large-scale mining will stop as deposits become less economical. To maintain the maximum benefit to the public interest, it is desirable to stop mining even sooner—as soon as the emergy yielded becomes less than that of inputs and environmental losses (Figure 6.5).

There is a long period when land productivity is lost during and after mining. Permits to mine ought to depend on the net emergy benefit of the mined products compared with the losses from interrupting land productivity (even when the mining is economical). Various subsidies for mining that were appropriate during growth—for example, allowing mining on public lands, and reducing taxes to accelerate mining (depletion allowance)—need to be removed. Laws that gave mining rights priority over surface use of lands are expected to change.

In order to make a public economy efficient, the mining, reuse, and release of materials need to make sense as part of the whole cycle. The free competitive market based on money considers only part of the cycle. Material cycles need to be operated for maximum public value measured with emergy.

Restoration and Reclamation

Where mining or other land use is extensive, regrowth of vegetation necessary to restore soil fertility is often delayed for many years for lack of prompt reseeding. Without reclamation, pits and gravel piles in gold-mining regions take more than 100 years to develop vegetation. Even with seed sources nearby, land productivity following Florida phosphate mining had recovered about 70 percent of its function after about seventy years. Much longer times are required in arid regions. Reclamation efforts in which vegetation is reseeded are usually a net emergy benefit. In general, more restoration is needed.

NEW SCIENCES OF ENVIRONMENTAL PARTNERSHIP

To better join the economy and the environment, new fields of science with new journals can be helpful in evaluating choices:

- *Ecological economics* revises neoclassical economics to include environmental causes and values (for example, using emdollars to choose alternative land uses).
- *Ecological engineering* combines environmental technology with self-designing ecosystems to make a prosperous interface between the economy and the environment.[4]
- *Environmental health* keeps environmental systems healthy so that humans are healthy (for example, protecting biodiversity).
- *Systems ecology* views the environment and economics as systems. Models and computer simulations are used to test understanding and anticipate the consequences of proposed changes.

• *Industrial ecology* studies the relationship of industries to the environment, considering more of the cycle of materials and manufacturing as a system.

SUMMARY

During descent, *assuming population decrease,* society can reorganize rural landscapes symbiotically with decentralizing cities. Natural capital can be restored. Environmental contributions to an economy based on renewable resources can be increased by rotating lands, reclaiming and reforesting bare lands, and sustaining complex forests and biodiversity. Less-intensive agriculture and forestry will require more lands and more labor. Recycling can replace most mining. If the priorities are made to maximize empower, green productivity landscapes can be refreshed for a better way of life.

17

TRANSMITTING KNOWLEDGE

A wave of information is spreading over the world, changing cultures and ways of life. Because of the explosive increase of computers and technologies in recent decades, better communication is unifying varied peoples, making their industries more efficient and generating faith in the future. We think information can do miracles without end, but there are limits to the quantity of information that can be supported. This chapter considers how to transmit information and knowledge to the future for the time of descent.

Information is in the parts and relationships that make systems operate and can be extracted from the systems as instructions, maps, programs, descriptions, plans, and diagrams. Information includes the genes by which we inherit life from our parents. It includes the culture and technology, which are passed down to us by our schools or gained from experience. Useful information can construct, organize, and operate complex systems.

Society and its economy are based on both kinds of information—that in the genetic inheritance of life and that which is learned. The economy of the future needs the genetic information carried in human genes, the genetic information carried by environmental biodiversity, and the learned information transmitted through education. Learned information that is comprehensive and organized so as to provide understanding is called *knowledge.* Incorrect information about systems is *misinformation,* and its use interferes with useful functions.

LIMITS TO LEARNED INFORMATION AND KNOWLEDGE

Maintaining useful information takes resources (see pp. 71–72). Large emergy flows are required to develop useful information for the first time, since each element has to be tested, revised, and selected over

Courtesy Matt Wuerker for Common Courage Press.

and over until it and its system are operating successfully. Examples include engineering designs, books, computer programs, memories, and music disks.

Information, like other kinds of assets and structure, depreciates when the items that carry the information deteriorate. People forget;

computer disks get fungus; books decompose; genes in living cells die with their cells; errors develop when copies are made. In other words, information follows the second energy law of degradation like everything else.

Judging from the brain's neurons and computer microchips, information processing is faster when miniaturized. However, information storage may not last long on the tiny scale where it can be disturbed by the normal shaking motions of molecules.

Copying information–for example, making a copy of a letter; making many copies of CDs; trees making many copies of their genes in seeds–costs relatively little in the short run. In the long run, however, information is only sustained by operating a population of units in an information circle (Figure 4.6) continually selecting, copying, sharing, and testing.

For learned information, people have to be continuously trained, information storage protected, materials recopied, errors eliminated, and information equipment maintained. Extensive electric power is needed for much of this civilization's information processing.

Because of the large resource requirements for using and maintaining information, knowledge is at the top of the energy hierarchy (to the right in Figures 4.8 and 17.1, with high transformity). Landscapes' knowledge and information are concentrated in cities–the hierarchical centers (see Chapter 7 and Figure 7.7). Some regard the level of knowledge as an indicator of progress.

Emergy to find and use information goes up very quickly when more information is added because that addition increases complexity. Information in excess requires more and more resources because of the difficulty of identifying one item among so many. The cost of retrieving books in a library goes up faster than the number of books added.

Information can improve and increase efficiency, but it cannot replace agriculture, industry, and commerce. A society with only information jobs is not possible. More, not less, fuel is needed for the electric power plants to support increases in information processing. Economies using high levels of information have not become independent of resources.

INFORMATION SURGE AT THE MILLENNIUM

In 2001, information, and especially its technology, were still growing, spreading in a wave that will probably reach its limit after the crest in the growth of the economy (see Chapter 5). At that time the number of books, universities, computers, and highly trained people may reach its maximum.

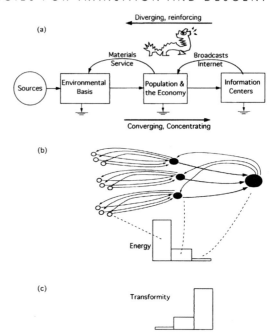

Figure 17.1. Information and society. The dragon represents an oriental symbolism for the power of public opinion. (a) Systems diagram of the basis for information centers; (b) convergence of energy transformations using up available energy and the return feedback of useful reinforcements; (c) increase of transformity toward the center.

As communication has become worldwide, it has made shared world values possible, especially through television (see Chapter 9). Examples are shared language, shared ideals, shared purposes, and shared images of authority. Shared information allows people to act in concert with great effect. Examples include disaster relief, military actions, and the public response to the Valdez oil spill in Alaska. Group consensus changes individual beliefs. It can settle conflicts. Sharing and transmitting information has been making more and more cities of the world—Miami, Taipei, Singapore, Tokyo, and Brunai, for ex-ample—into regional centers for processing knowledge, commerce, and finance. Shared communication helps replace transportation. Of course, much emergy is needed to copy information to millions of people and to get them to absorb and share it.

At this time, information is almost overwhelming in quantity and complexity. Many people think of information as cheap and thus easily substituted for other goods and services. It is a widely held belief that we are in a "post-industrial information society" (see Chapter 3). Many think that information and educated humans can replace industry and agricul-ture. This is attractive to those with faith in the unlimited capacity of humanity, and is perceived as a repeal of the dependence of the economy on resources. However, because information is dependent on electricity, which is dependent on fuel, its growth and spread is limited.

Information Excesses

While in transition, the present economy is in an information flood. Television and radio evolved a mix of entertainment, useful information, and misinformation that succeeds in gaining the attention of most people several hours a day, which is a huge emergy allocation. Society needs to decide how much of this entertainment is optimal. At present, education on a limited budget indirectly competes with television entertainment. What is needed is for some of the excess support of television entertainment to be shifted to education.

By 2001, the power of television had emerged as the information-organizing medium of the global society. With it came extensive emergy waste in excessive salaries, massive duplication, emphasis on entertainment, unnecessary advertising, and control by commercial markets. The democratic ideals of free speech were twisted into license to drown its citizens with destructive distractions of the normal family information processes. Especially for the young, television's hours of sex, violence, antisocial motives, false idols, and too much entertainment counteracted society's systems of education at home and school. For the adults time was drained away from sharing family care, community activities, and contributing to production. The commercial emphasis of U.S. television can be misleading to people of underdeveloped nations, who may mistake free market advertising goals for the priorities of democracy.

Society has been struggling to find a means of moderation. One idea is to limit dollar expenditures on advertising. Another is to support those stations that do not rely on advertising and have charters to control unproductive content. The limit to personal salaries (see p. 187) would help reduce the luxury emergy in television.

Information Storms in Journalism

The storing of public feelings about an issue sets up conditions for a storm of public attention and discourse arranged by the media (the store and pulse phenomenon from Chapter 5). Under intense competition for attention and profit, journalists seek new storms each day. Excess television funding could be causing too much competitive behavior.

Especially during times of transition from growth to descent, the public needs discussions of longer-range changes in progress and the principles responsible. In 2001, reporting was mostly short term; programs sought controversy rather than balance; reporters assigned motives with little basis and offered personal opinions as news. Television journalists confused entertainment with news and forced interviews into an adversarial win/lose sports format. Journalists tried to be the judge between adversaries. More emphasis was on gaining attention

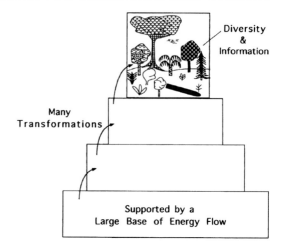

Figure 17.2. Diagram to show the high emergy of information

than on educating. TV commentators rarely questioned the free market growth ethic. Most of the media were oblivious to energy principles. Descent was not considered.

GENETIC INFORMATION AND BIODIVERSITY

Information in an ecosystem is in the DNA of the species' genes and in the complexity of organization of the species and environment. Part of the genetic information is in the organisms of ecosystems and part is domesticated in agriculture, forestry, and aquaculture. Humanity and the green cover of the Earth depend on this global network of genetic information in the species of plants, animals, and microorganisms. The most valuable is the variety of genetic information in humankind.

Biodiversity refers to the number of species, the legacy of past evolution. Biodiversity is a collective name for the inherited information of ecosystems, requiring large areas of support (Figure 17.2). The more kinds of life there are, the more complex is the ecosystem, and the more division of labor there is among the functions. On a larger scale, the more kinds of ecosystems and the more stages there are in areas in the region, the more diverse the landscape. Diversity is ready to supply suitable species when conditions change. Maintaining biodiversity has been difficult as natural areas are displaced for urban development and insatiable market demands for environment products.

Information Handling by Ecosystems

Nature's ecosystems give us perspectives on the use and limitations of information and growth (see Chapter 5). As a rapidly growing ecosystem begins to mature, it diversifies, develops division of labor, organizes, and increases efficiency. Information is developed at two levels as required to operate these higher levels of organization: (1) within genes in the organisms, and (2) in the structure of relationships of organisms and species. Information in nature, like knowledge in humans, is high in the hierarchy of energy processing (high transformity, Figure 17.2).

Studies of island species show a spatial limit to the number that can be supported. In laboratory ecosystem experiments, ecosystems with limited space and resources had an upper limit to the number of added species that could be supported. On the whole Earth, having large separate continents allowed evolution of different sets of species, increasing global diversity.

After information in ecosystems reaches a level for maximizing functions, additional information is put in reserve as a *gene pool* of alternates. For example, tropical rainforests where conditions are well-suited for life display a very high diversity of small populations of rarer birds and insects. Ecological systems operate most of their basic processes using only part of their stored information. Small populations of rarer species preserve excess information, keeping it in reserve as a gene pool available when changing conditions favor different species to become dominant. Gene pools are also a resource for economic domestication because of their information on special biochemistry usable in foods, medicine, biotechnology, and manufacturing.

INFORMATION DURING DESCENT

In ecosystems, ponds and forests prepare for the decreased resources of the winter by reducing operations and information in use. They save other information in the form of stored eggs, spores, seeds, capsules, and migrating or hibernating animals.

With economic descent comes less emergy, and information use may be expected to decrease too. Research budgets will be smaller and progress slower. Figure 5.6 shows the loss of high-transformity items when the base is less. Reducing population and infrastructure is not a loss of quality, but losing information is. Even as marvelous new technology for handling information is increasing, some basic information of our civilization is already being lost. Some of the fuel-based information of human society needs to be mothballed.

During our descent, as much species diversity and knowledge needs to be saved as possible, but for efficiency the information needs to be organized first. One tendency will be to give priority to saving knowledge in use at the time of descent. A different concept gives priority according to the effort previously invested in items of knowledge (according to transformity). What belongs on a priority list: principles, stand-alone summaries, textbooks, methods, popular classics, music, art, technology? The more widely information is shared on the Internet, the more likely that information will be preserved somewhere in the lower-energy times.

A main priority of society could be the packaging and storing of the part of the knowledge of civilization that is not in use. Preserving the wild genetic information will be especially difficult if the human population does not decrease fast enough, becomes destitute, and strips habitats. A species can be preserved with its ecosystem if the area is large enough for its population to operate a healthy information circle (Figure 4.6). Botanical gardens and zoos are another way to save genetic stocks for short periods, but the numbers are few and their future support will decrease also. Placing endangered species—such as ground parrots and a species with anatomical features of ancient extinct reptiles—on protected islands has been successful in New Zealand.

EDUCATION

As everyone knows, education transmits society's information to children and adults for the future. Essential information has to be duplicated and taught so that it is shared by each rising generation. A human use is required for extracted knowledge (encoded information isolated from its system—see p. 71) to be functional. If information becomes limited, choices will have to be made as to what is taught. Ways of sustaining motivation might be the most important information for a time of change.

During growth, emphasis was on getting new information. There were many creative experiments with curricula. But as resource availability declines, emphasis is on efficiency in teaching information that we already have. There is already less total money for administrative salaries, buildings, libraries, teachers, materials, and state and federal education agencies. Progress in education may have to be made without much increase of budgets.

Efficiency with fewer resources means reducing bureaucracy, reducing salaries of principals and superintendents, eliminating paperwork, cutting nonacademic activities, stopping outside interruptions, and not tolerating misconduct. It need not mean increasing class sizes and conformity to central control. Removing problem children to alternative ways of learning can improve the teacher-student atmosphere in schoolrooms. One alternative is for uncooperative students to learn motivation by participating in groups working on community projects.

Efficient physical education means less expense on spectator sports and more participation by all students. Because of equipment costs, sports with few equipment requirements like soccer and touch football may replace more expensive sports. Let's replace high-energy interschool sports teams–oriented toward championships and a few stars–with intraschool sports that include all students.

New initiatives such as distance learning, magnet schools, home schooling, vouchers, and charter schools need to be evaluated for their teaching efficiency and emergy benefit along with the older kinds of schooling. Teaching courses to dispersed people with the Internet and e-mail is flexible and saves costs of transportation. Whether the proven benefits of learning together as a group can be achieved at a distance remains to be seen.

However, in a time of decentralization, it will be important to give all students a common core of knowledge: culture, civics, science, and arts. This uniformity can be achieved best with public schools. Jurisdictions, laws, taxes, and budgets should be organized to give a state-approved standard education to all children including those in private schools. One suggestion is that about half of what is taught be according to standards of federal, state, and local authorities, with the goal of sharing common knowledge and achieving basic skills. The other half might be left to teachers to generate innovations, creativity, and initiatives.

Ways to cut can be different in each school district, part of trial and error for descent. We already have some experience with video-tape teaching, computers set for self-teaching, and interning for vocational learning. Other schools will copy those choices that work.

An economy of effort might be made by reducing schools from twelve grades to eleven, as they were in many places in the United States early in the last century. Students would move into college or jobs sooner, possibly decreasing problems with idleness, teen-age diversions, drugs, and precocious emphasis on sex. Summers can be used more efficiently for make-up courses, special programs, and job and intern experiences, offering organized opportunities for every student.

Courses

Although schools have offered combinations of many courses to explain our complicated world, students rarely understand how these pieces fit into their future lives or how to deal with problems. A new course is needed that explains how the three E's are related: *energy, environment,* and *economics.* The course could use systems ideas, simple computer simulations, and emergy concepts.[1]

Since many of the specialties needed in the present era of transition are changing, narrowly educated people have to be retrained for new kinds of jobs. Thus, it is not appropriate with tightening budgets to teach only traditional subjects (reading, writing, arithmetic, classic science). Broad generalist training should also deal with the larger scale of environmental science and economy. Teaching should show how the principles apply to more than one scale of knowledge. For example, the same principles of growth and competition apply to ecology, economics, chemical reactions, anthropology, and fisheries.

Today we teach and test students as isolated competing individuals. A new trend is to teach groups in a cooperative and participation mode more like the situation in their future jobs. Local examples make overview studies real and interesting. For example, a rural school in a farming area could emphasize low-energy agricultural systems and policies. In a city area, examples can include that city's problems adapting its industries.

In 2000, many high school and community college programs had specific vocational goals. Not everyone was expected to go to the university. While students were still required to take basic courses in English, math, and science, they spent some of their time and energy learning specific job-related skills. Since these students chose these tracts, their motivation was high. However, in times of transition and descent, keeping job training up to date may be difficult. Part-time apprenticeship programs with local employers can help.

With fathers and mothers both working, preschool day care is often necessary. Head start programs for underprivileged preschool children are successful, but to reach all children, day care centers with educational objectives could be administered as part of the school system. This policy recognizes that *all children are the responsibility of all of society* and increases the productivity of parents. Former First Lady Hillary Rodham Clinton made the concept famous with the phrase *it takes a village to raise a child.*[2]

By selecting large books with too much detail, central selection of textbooks often raises costs and interferes with teaching principles simply. A selection of several alternate texts is better for a time of change.

Even with less travel, multilingual people are needed for trade, international projects, international sources of information, and representing foreign companies. Although immigration pressures should decrease later when more of the world is more equally developed and populations are smaller, communities speaking foreign languages will still be a part of the time of transition. Retaining foreign language as a basic education requirement keeps doors open for further training toward an international career.

Internationalizing the curriculum with multicultural education is a popular educational goal. Students with roots in other countries need to feel that their cultures are valuable. Bilingual people are a strength

Reprinted with permission of King Features Syndicate.

for a nation and its international relationships. However, a common core of shared information unites a pluralistic society. An official language should be maintained for common national values as well as an efficient economy.

The goal of bilingual education is to prepare students as quickly as possible to speak and learn in the main language. Allowing a second language to become an alternate language for their whole education keeps those students a disadvantaged minority and weakens national unity. Adaptation for those of different languages, with the goal of having them taught the nation's primary language, can be aided by preschool orientation and help in school. Younger children learn rapidly from friends. Teenagers may already have studied and learned the class lessons in a different language and can be helped with the transition.

Efforts are underway almost everywhere to make computers more generally available in schoolrooms and to develop teaching methods that use them effectively to aid learning. Computers are already a part of most jobs, so being adept at a variety of computer uses is becoming essential for most employment. However, during descent there will be other kinds of jobs developing on farms and local industries.

Because many families do not have computers at home, getting computers at school for everyone is necessary for equal educational opportunity. The Internet provides extensive knowledge resources especially valuable for schools without much library access. Communicating over the Internet can help motivate learning. Rural people will be better integrated into mainstream work and activities by using the computer network.

An efficient economy puts its disabled people to work in ways appropriate to their abilities. Special education can be adjusted toward what is practical for making every citizen a contributor, even though some contribute more than others do. Including students with learning disabilities in the regular classroom (known as *mainstreaming*) is acceptable if special education teachers or their aides can help. The practice should not interfere with general classroom success or take teaching efforts away from the normal students with poor backgrounds. Life in special work institutions, such as diversified farms, is better for some than live-alone independence. Apprenticeship education may be practical for this goal.

With decreasing resources for transportation, more small neighborhood schools are likely to develop again, while electronic communication will be used more. With people dispersed, some busing will continue. Supplementing budgets is a way of equalizing education for students in various neighborhoods. Neighborhood schools can again be community centers.

An important educational goal is to help children grow up with classmates and friends who represent our national racial and ethnic mosaic. In the changing times of the 1960s and 1970s, because housing was mostly racially segregated, schools were looked to as the place to teach children to live with children from other backgrounds. In the absence of busing, new approaches are needed to foster diversity in the schools.

THE GREAT UNIVERSITY

A university president usually represents his institution as having three functions:

- *Teaching*
- *Research*
- *Service*

A fourth function may be the most important, but it is rarely mentioned to society, legislatures, foundations, or students:

- *Intellectual Leadership for the Long Run*

The campus is where most of the new and old ideas, methods, and technologies for progress and preservation are tried out in a potpourri of publications, discussion, late laboratory hours, disagreements, discovery, and new thinking, often contradictory. Later, governments and industry select from the university products whatever is pertinent to their need at that time.

Much of society's information is in the great universities, its libraries, and its scholar-teachers. As explained with Figure 4.6, useful information is sustained or increased only by continual processing through an information circle of testing, selecting, copying, duplicating, and reapplying. Colleges routinely help transmit to students the knowledge that is in general use. In addition, the universities sustain knowledge that is not in much use. A scholar can be one of a handful in the whole world keeping specialty information available. The practice of specialists keeping each line of information alive as a matter of individual pride, as an art form, or as an avocation is analogous to the gene pool's saving of genetic information. Items of unused information become critically important to use as conditions change.

Knowledge is threatened because downsizing has already reached universities as budgetary cutbacks of people and libraries. Faculty time has been moved away from the scholarly information pool on which the long-run economy depends and reallocated to more and larger classes. Some knowledge is carried in governmental and private research laboratories, but these have also been cut. The flurry of emphasis on computer data banks is risky because computer hardware's long-range storage abilities have not been proven, and because raw text is not useful without the scholars who can retrieve and explain it.

Whereas universities during the last century of growth have become part of the mainstream of currently relevant knowledge, development, and education, during descent they will have an increased role as repositories of *inactive knowledge.* Biotic information such as that in specialized agricultural varieties needs live preservation, which is costly (see p. 243). Whether keeping genetic varieties and knowledge for the future uses more resources than redeveloping them again later when conditions warrant depends on the time between uses.

Individuals, university task forces within the faculty, and students—particularly in a time of cresting and descent—ought to be organizing and condensing knowledge developed in the forty-year information explosion, writing summary books, and teaching the next generation how to draw new principles from the excess of already-published information. Perhaps committees could reorganize universities for these missions. Because the great university's most important mission is knowledge for the long-range future, university appropriations should not be limited to short-range purposes.

Difficulties and Efficiency Measures

For a time of downsizing, some universities are badly managed with funds for necessary primary functions draining away into excessive administrative salaries, nonacademic student activities, and bureaucracies that fill out unnecessary forms. Presidents and deans are often isolated from faculties, catering to student popularity, expanding student bodies to please developers, spending time on football alumni, seeking public popularity, judging faculty by notoriety and money brought in, taking the legislature's side on issues that show contempt for faculty, and ignoring real scholarship. Student grades, instead of being kept tactfully secret, ought to be posted and published and the stars publicly praised just as those in athletics. The right kind of downsizing would let some of the short-range projects go, shorten the personnel hierarchy, eliminate students who are not serious about their education, and redefine university goals.

In 2000, faculties were stressed and diverted from their mission. Too many intellectuals had been trained, leading to heavy competition for jobs. This keeps salaries low. Teaching loads are returning to their higher, pre-1950 levels, sometimes with huge classes. Faculties are pressured to do short-range work because it brings in outside money. At a time when little grant money is available for new ideas, young faculty members spend their most productive years writing grant proposals with nothing intellectual to show for it. Then they can be denied tenure for neither raising money nor developing a career avenue of original thinking.

Temporary part-time faculty without benefits are being substituted for regular faculty, and tenure is being eliminated in some places to reduce salaries and to allow release of faculty who have lost the zeal for teaching and research on their subject. This approach gets elementary subjects taught but does not maintain the knowledge base of society.

A better plan for selecting faculty dedicated to knowledge and teaching is to lengthen the advancement times (similar to those before the Second World War). People would be hired at instructor level, promoted to assistant professor level after three or four years, promoted to associate professor after another seven or eight years, and promoted to professor and given tenure after another seven or eight years (without any "up or out" rules). Budget limits should be solved with across-the-board cuts for everyone instead of using part-time faculty without benefits.

There is a great difference between the rich and poor students on campus. Students with little money must use up much of their time and energy at jobs, leaving little time to study. Many students without wealthy parents are being forced out by universities that raise tuitions instead of economizing. Often students are forced to pile up large debts, which are difficult to pay back. Society ought to be investing in its future by subsidizing students, as is done in leading countries of

Europe. To equalize opportunities, a few states are beginning to provide tuition for students with good grades.

For more efficient scheduling, quarterly terms could be used in place of semesters because they are more flexible, make good use of summers, get students through quicker, keep teachers from overemphasizing any one course, and give students more breadth, a property needed during times of change. University schedules can use more days of the week, as they did in the past. Classes can be six days a week, with Saturday morning for classes and athletic or social events in the afternoon. Schedules for typical three-hour courses can be Tuesday-Thursday-Saturday as well as Monday-Wednesday-Friday, using building space better. Libraries would be used more on weekends. To help new students get started with good learning habits, joining social fraternities and sororities can be prohibited in the first semester. Class attendance can be monitored for freshmen and their class cuts limited.

Traffic congestion that makes faculty and staff lose work time is caused by commuters parking on campus. Dormitories without parking spaces on campus ought to replace the campus parking garages required for commuting off-campus students. With fewer students going home on shorter weekends, and better bus transportation, student cars are not needed and could be kept off campus or prohibited. Unnecessary cars are a large emergy waste (Figure 12.3). In later years, with fewer students and cars, garages will not be needed.

The attention of society is needed to narrow the university's primary missions while downsizing. The university's job is not to seek consensus or popularity. It is not for short-range research for industry. It is not for students to be anti-intellectual, emphasize fraternity and sorority activities, or facilitate use of drugs, alcohol, and sex. It is not a place for emphasizing patents and profits. It is a place for long-range thinking, generation of new concepts and inventions, and consolidation of old knowledge for the time ahead.

Physical Education and Spectator Sports

Required physical education for all students provides regular exercise for good health, exercise habits for later life, and team spirit from group sports. However, using scarce budgets to emphasize interschool and intercollegiate sports often leaves the majority of students out.

High schools and universities have stumbled into being spectator sports entertainment centers, with millions of dollars made available from television. Sports events attract public interest to schools and universities, but they create the wrong image. Ignorance and even contempt of the primary intellectual mission is fostered when the coach is paid twenty times the faculty member and the commencement speaker talks about football. Players receive subsistence-level pay as "athletic scholarships" while the athletic department receives millions of dollars.

Universities could transfer the sports operations to private athletic clubs, as is done in other countries. The players would be paid as professionals with the option to register for courses and degrees like any student. The sports club could still carry the traditional name and symbols of the university and rent the university stadium. The university gets out of recruiting and forcing some uninterested players to study. Then the athletic departments can concentrate on physical education for all students.

SUMMARY

Part of global information is in the intellectual knowledge of society, and part is biotic information in the genes of humans and the Earth's life-support biodiversity. Sustaining the biotic information depends on maintaining large areas of diverse ecosystems through critical times of transition when human populations exceed resources. Because of the wealth used to develop and maintain information, its transformities are high and there are resource limits to the amount of genetic and knowledge information that can be sustained. Only part of our information will be used by our civilization in descent.

For adaptability and long-range progress, we also need to preserve unused knowledge. To maintain knowledge, institutions for education, communication, and technology require large amounts of electrical power. Information processing by television helps develop a globally shared culture and peace ethic, but to be efficient, society has to set television priorities among serious purpose, advertisement, and entertainment.

Education is the transmitter of knowledge. Table 17.1 summarizes suggestions for education during descent. With the right kind of downsizing, nonacademic waste can be eliminated and the university's intellectual functions emphasized. The great university, although smaller, may be the best hope for *leading descent and preserving knowledge from the era of growth.*

Table 17.1—Suggestions for Educational Efficiency

Reorganize jurisdictions, laws, taxes, and budgets to give a state-approved standard education to all children. Require this standard for private schools also.

Provide day care for small children, extending through normal working hours as an automatic part of all schools.

Provide preschool orientation and help in school to those of different languages with the goal that they can be taught in the nation's primary language without handicap.

Provide after-school activities for all ages.

Replace high-energy interschool sports teams—oriented toward championships and a few stars—with intraschool sports that include all students.

Provide a unified systems course in energy, economics, and environment taught by teachers trained to relate these fields to global change.

Let about half of what is taught be according to standards of federal, state, and local authorities, with the goal of sharing common knowledge and achieving basic skills. Let the other half be left for teachers to generate innovations, creativity, and progress.

Encourage use of new technology like computers and the Internet, experimenting with various new ways of using these in teaching.

Organize school schedules to occupy students in learning through the whole day and year-round. Learning can include after-class projects, internships, and vocational experiences making use of community help.

Let schools supplement medical care, nutrition, and job placement for students without these at home. Use parent-teacher activities to get more community support of all the children.

Organize district-wide programs for racial and ethnic variety in schools.

18

PREPARING PEOPLE

In this chapter we try to anticipate what descent will mean for people's lives on the smaller scale. If the future is to be prosperous, people will need new ideals, extended family patterns, changed lifestyles, and a cooperative attitude to accept change.

IDEALS FOR NEW CONDITIONS

For transition and descent, new personal values and perceptions of progress should emerge. The "growth is progress" ethic of the industrially expanding world is being replaced now by "sustainability is good" for a world that is no longer growing much. Next, as people learn to live with descent, the ideal will be "down is better."

During growth, many individuals seek money, political power, or celebrity. These characteristics help the economic system maximize empower when there are new resources to develop. But during descent and later, a different happiness principle is needed:

> An individual will feel a sense of worth by contributing to the larger system of environment and society.

The values for a growing economy are larger size, greater speed, and competition. Values for coming down are smaller size, efficiency, and cooperation. Faith that public good comes from individuals greedily seeking personal gain in a competitive marketplace should fade with the other ways that were useful for accelerating growth.

The pioneer social ecologist W. C. Allee, comparing animal and human societies, used the term *cooperative-competition* for mild competition of individuals that is beneficial to the whole group.[1] For example, people on a farm made a game to see who could harvest the most vegetables. Beyond growth, cooperative competition replaces the destructive kind of competition to eliminate rivals.

Newspaper columnists and talk show hosts often sense changes in values; for example, the columnist Ellen Goodman wrote that the motto of the 1980s, "You can never be too rich," has been thrown out. To be too rich has now been dubbed *greedy*. She wrote that spending is out and *service and sacrifice are in.*

For many people, religion provides a basic code of behavior and a way of teaching morality to their children. In times of confusion and change, many people retreat to their traditional doctrines, which may need to change on some issues for the new times (such as the emphasis on reproducing more children). Religions adapt to reality, but slowly, because of the inertia of millions of people over wide areas sharing common beliefs. Notice how slowly the attitudes of many clerics are changing on population growth, environment, war, and women's roles.

Religions classically were concerned primarily with human relations on the small scale that children could understand, but were mostly silent on large-scale issues such as environment, energy, capitalism, trade, and international organization. Advocating small-scale fundamentalism at this time may be distracting if emergy-based morality is needed to indicate what is *right* on the large scale. Large-scale ethics could become a greater part of Sunday school religious teachings for adults.

MECHANISMS OF SOCIAL CHANGE

The pulsing paradigm principle from Chapter 5 may explain how a society accepts new ways: *It accumulates gradually the means for subsequent sudden action.* As change occurs, individuals begin to think about what is different and a general shared feeling builds up gradually in the society. Then at some threshold the new idea jumps into the forefront of social discourse, and most people's attitudes flip together. If people were asked individually by a pollster before the collective flip-flop, most people would express the old consensus. In other words, there is a gradual buildup of thought followed by a public discussion and new consensus. For example, in this century the public opinion in the United States about an appropriate number of children has decreased from four or more to two or fewer.

Although individuals try various alternatives, the reality of what works ultimately determines group policies. Decisions found to be best for the whole society become the new accepted pattern, but individual differences are a contribution to the process of finding what works. The flexibility of this adaptation mechanism keeps society tracking its basis of support. It is one reason why democracies can be expected to adapt quickly and lead the descent.

FAMILIES

Self-organization of human societies usually develops around the family. Figure 18.1 shows the main connections between the family unit and society. The family receives direct inputs from the environment, purchased inputs based on the money gained from work, and inputs from government for education and services. The money (dashed line) received from work is paid out for necessities and taxes. With fewer urban resources, families shift to the left in the energy hierarchy (Figure 5.6). Families receive more from their local environment. Government tax and welfare support decreases. At the beginning of the twenty-first century many families and individuals are already struggling with diminished real wealth.

Everyone needs regular social exchanges with others, and the family often supplies part of these. Maintaining some kind of family organization seems to be the most effective way for individuals to meet their needs and for society to be productive. Family ties foster feelings of closeness, security, and caring that people need to be successful.

But during descent, genetic families will need to be smaller. Thus other associations, such as extended relatives or friends living together, may develop. When people join together into extended families they will be more likely to share than to pay for baby-sitting or a ride to the grocery store. For stability, the tax laws that benefit families as the orthodox lifestyle will need to be broadened to accept more options for organized relationships.

Within the extended family, some hierarchical relationships are expected. (Hierarchy is a natural pattern—see pp. 65–71.) Equity within the family group is also necessary to help each person contribute his or her best. This kind of equity means fairness, not equal position in the control hierarchy. Each person needs to feel his or her abilities are respected. For children, their positions in the hierarchy usually go with their age (and thus education and experience). For adults there can be

Figure 18.1. Energy systems diagram of the family and its main inputs and outputs. Dashed lines show the circulation of money.

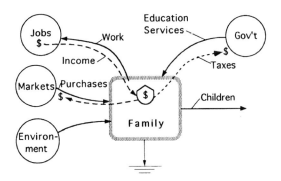

agreed-on control according to what is needed, recognizing many reasons such as knowledge, experience, time, and willingness to accept that responsibility. Over 60 percent of U.S. women in the year 2000 were working outside the home, many with arrangements for sharing housework and child care, including hired help.

Children

Deciding to have fewer children is already a trend that keeps a higher per capita emergy budget for the family members. Some realities of the lower-energy world will be hard to accept, like not expecting your children to have a life richer in consumer goods than you have had. Children may live at home longer or move back home after having trouble getting started on their own.

It is no longer frowned on to decide to have no children. Many couples see advantages in having two careers and no children. As growth has leveled, not so many children are needed and they are sometimes an economic liability. For many parents, particularly single parents, children seem a burden. Child abuse has increased, requiring society's help for both parents and children. In principle, greater attention can be given to each child if there are fewer children.

Women can have fewer children and more education for longer careers of productive employment. Societies making these changes promptly will prevail because of their greater efficiency. Especially in poorer nations, population control can be achieved not only through the use of birth control devices but also by giving women the rights to property, employment, and the means to start small businesses.

Housing

Once society corrects the excessive salaries and unearned flow of money and emergy to the rich (see Chapter 13), the housing industry can build smaller, more efficient units available to those with ordinary incomes. The United States now has extensive little-used housing in second homes, excessive tourist facilities, and unused rooms in luxury residences. With lower incomes and less money available for housing, much of this excess can come into general use. Many larger homes can be converted into duplex arrangements or small families can live together. More durable housing structures will gradually develop (see pp. 214–215).

Personal living space will decrease. More people will live together in the older houses. Second vacation houses will be hard to keep. Older, energy-saving architectural designs (such as passive solar technology) can be used that take advantage of nature, like planting deciduous trees for summer shade and winter sun. The name *xeriscaping* is sometimes given for native styles of landscaping that are not so dependent on water and fertilizer. If forest management provides wood on a renewable cycle, houses with fireplaces and wood-burning stoves for auxiliary heating can become useful, desirable, and stylish. If people are more dispersed, the wood smoke and ash can go back to the lands as a fertilizer rather than becoming an air pollution hazard.

EMERGY OF LIFESTYLE

Choices for the good of the family and individuals need to maximize real wealth (emergy use), which may be different from income. People moving out of the city may increase their real wealth even though their income decreases. A person on a farm draws many emergy values directly from the lands and waters, using money less than in the city.

If during descent population is not reduced fast enough, the emergy per person decreases. Even when the economy is level, the steady increase in workers reduces the average income of each. People need to understand that this loss of income is not their personal fault, but rather caused by the increase in population and the decline of world resources. With less emergy per person (less buying power), many people will have to reduce their consumption. Most will have to make difficult choices about what to do without. They may choose fewer or smaller cars, smaller houses, less-elaborate vacation trips, more economical clothes, or fewer special programs to enrich the experiences of their children.

Values of family work are underestimated because much of family work is unpaid. When we pay for day care, housework, and yard work, we are putting market values on some family functions, but the values of caring, loving, consideration, and mutual support are a kind of high-emergy information exchange not measured with money.

INDIVIDUALS

During descent as now, each individual needs several relationships with society. Policies described in previous chapters can help provide these needs.

Jobs

A minimum-wage public works program (see Chapter 12) and substitution of salary reduction for firing (see Chapter 13) should ensure individual employment. With a job comes basic comforts.

As explained with Figure 13.1, emergy and income from jobs are distributed according to energy hierarchy, with many low-level jobs and few high-level jobs. The more able, educated, and knowledgeable you are, the more likely it is that you can get one of the jobs at the right end of the hierarchy (those that require higher transformity). Then when the total emergy of the economy decreases,

THE FAMILY CIRCUS by Bil Keane. Reprinted with permission of King Features Syndicate.

the upper-level jobs tend to disappear and people move toward the left.

Family Group

With more flexible attitudes toward more kinds of family groups, most individuals can join some kind of extended family (see pp. 271–272).

Health Care System

Whereas the free market health care pattern has extreme expenditures for some and none for others, the baseline health care system (see Chapter 12) can protect all individuals. Already a part of social change is a new sense of individual responsibility for health: eating less fat, practicing safe sex, and doing more exercise by walking and riding bicycles.

Efficient Lifestyle

With more responsible attitudes that come from understanding how individual waste detracts from everyone's essentials (Figure 12.1), individuals can select a more efficient lifestyle.

If you can live close to your job, perhaps you can trade one car for a bike. Society begins to admire those who own small cars and make them last. As cities decentralize into smaller communities (see Chapter 14), with jobs closer to home, there is an expectation that people will get to know their neighbors better and communities will become more cohesive.

Large-Scale Purpose in which People Can Have Faith

If descent becomes a recognized mission by schools, churches, and families, these groups can help individuals find inner satisfaction in contributing to this common purpose.

Sex Outlet and Reproduction

Over the centuries in which human society increased its population and role in the planet, races and their cultures were selected for their high reproductive potentials. In recent times of growth, procreation was the predominant emphasis in church and family. Now that less reproduction is needed, can our species with its biological drives and cultural emphasis on sex de-energize reproduction so that overpopulation will not cause a crashing descent?

It might be predicted in principle that alternate human relationships, those that can absorb sexual energy *without reproduction or disease,* will emerge and become accepted. The strongly inherited sex drives of the human species can be coupled to behavior in many ways, depending on inherited tendencies, cultural training of children, and attitudes of adolescent peer groups. In the 1990s we observed trial-and-error experimentation and public discussion of various practices for control of reproduction and/or sexual energy. These included same-sex pairing, birth control, legal abortions, abstinence, masturbation, unconsummated sex, delayed marriages, marriage without children, marriages with adoptions, more control of teenage sex, and struggles against religious dogmas opposing changes. In principle, practices that reduce population will be reinforced and predominate during descent (see Chapters 8 and 12).

Mental Health

The decreased intensity of urban life and redevelopment of a less-stressful rural life may result in less mental illness. More mentally handicapped people can find useful roles. As part of the priority for preventative medicine, society may be expected to continue its policies of supplying the drugs that stabilize many inherited mental illnesses.

Some explain the epidemic of harmful drug use in our society as a diversion needed by people without a vision or shared purpose for the future. If so, having everyone join a collective mission for prosperous descent could help many find social purpose and a more zealous life.

PLURALISTIC HARMONY

Economic benefits are likely to help cultures that use their diversity to help productivity. Beneficial pluralism may spread. Information-sharing policies were suggested for peace among peoples of other nations (see Chapter 9).

In descent, two contrasting tendencies will mold public attitudes on human diversity: (a) global television develops a sense of belonging shared worldwide; (b) people develop local values as they move out of cities to more isolated small communities. The diversity of local culture can increase again but with aspects of world consensus and mutual respect. Politics, economics, behavior, dress, music, architecture, and art can be unique locally but still, through television, interact with and be influenced by shared world culture.

The new organization of the global society seems to be overriding the antagonistic, warring attitudes that were acceptable in earlier times of competing territories. Global television and business communication have joined many of the peoples of the world to think similarly and share purposes. Positive attitudes are forming toward people of different races, cultures, and creeds.

During early colonization and growth in previous centuries, the mechanisms of fast growth of the fuel-consumer economy simply overran those that were different. As in the weedy, colonization stage of ecosystems, uniformity with the fast-growing dominants was regarded as part of progress. For example, on the American frontier, minorities—both the Native Americans and the minority colonists—were not tolerated. Now, with the cooperative ethics of a climax society, reintegrating all peoples for the common good is regarded as progress. Obtaining respect and benefits from different cultural roots is the strength of plurality, aided by the knowledge of past cultures gained from anthropology.

Considerable energy may be lost in trial and error, but more creative choices and initiatives may tend to keep open societies adapting and productive. Today's inefficiency of new complexity may be tomorrow's innovation and progress. Stress during transition may be less if we understand and plan. When there is less population and economic activity, times may be more tranquil.

Greater empower comes to all when individuals are allowed to find their own best ways of contributing to society. Immigrants and descendants of original native peoples can find their own balance between old cultural ways and those of the dominant society in which they are now included. Public policies should not to try to rewrite and reverse history about minorities, but they can arrange whatever special subsidies are necessary to give individuals appropriate opportunities to serve.

SUMMARY

If they understand the need, people can adopt the concepts of prosperous descent, adapt their lives, and regain zeal for the future. Assum-

FOR BETTER OR FOR WORSE by Lynn Johnston. Distributed by United Feature Syndicate. Reprinted by permission.

ing some success in reducing population, moderating income extremes, and restoring environmental life support, people can find interesting and meaningful lives without much reduction in real wealth per person. People during descent may live with more stable employment, more varied family groupings, fewer children but better child care, smaller housing allocations, more aspects of their lives in local interaction, tolerance sustained by global television, and new roles as custodians of culture and information.

19

SUMMARY FOR ACTION

This book uses principles that apply to all systems to anticipate the future of the human economy on planet Earth. From these principles, policies are proposed to make smooth our current time of transition and make prosperous the contraction of our civilization just ahead. Knowledge about the analogous ecological systems helps explain the principles of growth and descent. That ecosystems can descend in orderly steps means that a prosperous turndown of global civilization is possible.

Most people don't believe that human affairs are determined by scientific principles. Because humans are embedded in the system of humanity and environment and see so much detail, they don't realize that the human system in aggregate is following the same laws of energy, materials, and information that apply to all scales of the universe from molecules to the stars. For lack of understanding of the noisy complexity of our world, people pursue beliefs in randomness, indeterminacy, miracles, money, magic, predestination, divination, anarchy, and selfhood. Instead, in this book we offer simple systems principles to explain our existence and what to expect next.

By understanding how people and their cultures evolve by principle, an individual can see what is successful, what is right, what is worthwhile. The individual can adopt goals for a life's dedication that are satisfying because they are systems worthy. By sharing these beliefs, a unified society can take the necessary steps of progress, heretofore achieved only by wasteful trial and error. For global humanity to plan ahead together would be a giant leap forward.

For six billion people deeply indoctrinated in the ethics of growth, a turndown and descent of civilization is unthinkable. That this descent could be prosperous is so inconceivable that it is unmentionable. This

HAGAR THE HORRIBLE by Chris Browne. Reprinted with permission of King Features Syndicate.

book makes a start on opening the discussion. Showing a good way down is a call for everyone to think ahead and plan.

The Present Condition

We started by looking at our present condition and other people's views of it. Indices of economy, population, and environment at the end of the twentieth century show the growth regime of several centuries to be faltering. But the short-term ups and downs of politics, economics, benefits, disasters, and human misery are so sharp that they mask the long-range trends.

We reviewed other books about global futures. The authors agreed that drastic, accelerating changes were in process, but disagreed about the kind of future to expect and what the policies should be. For lack of a clear vision about the future, there is confusion and dissension worldwide.

Some of these authors reviewed the history of civilizations in search of repeating patterns to explain current times and trends, but our times are unique in the size of the energy resources involved and the comprehensive power of the global sharing of information. Some sought causes in people's behavior, but that scale is too detailed to synthesize an overview. Instead, this book applies principles to a larger scale to understand and anticipate trends in the aggregate.

Systems Principles

Our general systems principles include concepts of energy and materials, universal energy hierarchy, emergy and transformity, maximum empower, emdollars, cycles, money-emergy relationships, pulsing, spatial convergence and divergence, and population regulation.

By explaining the universal energy hierarchy, energies of different kinds are shown to be different in their ability to do work for the economy and environment. Energies of different kinds are made comparable by expressing each in terms of the energy of one kind that was previously required for their formation. This quantity, called *emergy*–spelled with an "m"–is a general measure of *real wealth*.

The principle of *maximizing production and use of real wealth* guides the self-organization of systems on all scales, including the realm of the human economy. This is the *maximum empower* principle used in this book to suggest policy. Production by successful designs supports consumer units that feed back their services to reinforce production.

The more work is applied in successive steps, the higher is the product on the scale of real wealth value. *Transformity* measures the position of something in the series of successive steps toward higher and higher value. This universal scale of value ranges from the dilute solar energy at the low-transformity end to genes and shared information at the high end. The *emdollar,* the monetary equivalent of emergy, puts resources, components of the environment, products of the economy, human assets, and intangible contributions of humanity on a common basis. *Maximizing emdollars* is the basis for choices among alternatives in the public interest.

The closed circulation of materials is described as a necessary design for sustaining any system. Using the energy of the system, any cycle of materials converges from a dilute state in one part of the cycle to a more concentrated state before recycling. Fitting the principles of cycles is the basis needed for managing materials in industry and the environment.

All systems prevail by regular pulsing. Universal oscillations are in all scales, from elementary physical particles to galaxies. The mechanisms are similar. As energy is processed, products accumulate gradually. This stage is followed by a surge of use and growth by a consuming activity. Small scales pulse rapidly; large scales slowly. The dynamic pulsing on all scales at the same time produces the noisy fluctuations we see in the real world and often misinterpret as randomness.

One of the pulsing patterns is the system of humanity and its Earth basis. The several centuries of growth of our civilization on the previous accumulations of fuels and other resource reserves is recognized as one of four stages in the pulsing cycle of global civilization. Whereas a steady state is not possible, our human system is sustainable in the long run if we learn to adapt to the appropriate stage in the alternating sequence of growth and descent. The present stage is *transition,* and *descent* is next.

In describing the economy we show that the circulation of money fits energy systems principles. The free market mechanisms add efficiency. The money-emergy relationships were shown for the U.S. economy for 1997. Emergy-money ratios for different countries show differences in the buying power of national economies. Comparisons among countries show wide differences in real wealth per person. Because money is paid only to people and not to the environment, most purchases of environmental commodities yield more real wealth than is paid for them. Emergy indices are necessary to make choices in the public interest. One ratio evaluates energy sources to see which make a net contribution to the economy. Another ratio measures economic-environment relationships to decide on the feasibility of developments and their environmental impacts.

All systems develop spatial organization according to the principles of distribution of energy and materials. Spatial organization is hierarchical, with peripheral areas converging to centers of concentration. On land, stream branches converge to the river mouth, and geologic processes and precipitation converge in mountains. Human settlements develop where these energies are most concentrated. In our fossil fuel era the cities are also centers of fuel use, with a hierarchical pattern of night lights vivid to satellite view. Circulation of money, concentrations of emergy use, and information processing are organized according to the distance from centers. Indices of money and emergy provide guidelines for planning and adapting human society to the landscape for maximum mutual benefit.

The work of the human population generates real wealth, and real wealth is required to support humans. The real wealth per person is already diminishing. The effect of increasing people on global productivity has a diminishing return, especially as available fuels diminish. More kinds of sexual outlets that do not lead to reproduction or disease will become customary. As the critical importance of reducing population becomes general policy, many kinds of birth control will be accepted. A global simulation model shows that an abrupt turnaround from population growth to sharp decrease may occur if support for public health and medicine declines. The way people depend on environmental carrying capacity is dramatically demonstrated in Biosphere 2, the glass mesocosm in Arizona.

Policies for the Present Transition

We apply the systems principles to the current time of transition at the end of growth to explain what is happening and to suggest policies for international organization, energy use, the national economy, and making people productive.

To balance global wealth and eliminate affluence and waste, priorities are needed to develop real wealth equity in international exchange and control global capitalism with such measures as an emergy-based worldwide minimum wage. Mutual and free sharing of information is economically beneficial and aided by the new communication technology, television, and the Internet. Aggressive educational programs can generate a globally shared attitude that agrees on the means for international peaceful behavior and pluralistic respect, while retaining pride in local differences in language, custom, and cultures. Emergy evaluations indicate the real power of nations and a way to assign votes in international organizations. An emerging policy of international control of resources by shared military force can help make nonrenewable fuels available to all countries through the open market. With shared international attitudes on global relations, agreements can reduce the needs and costs of military defense establishments.

Net emergy evaluation of alternative energy sources show the fallacies in many alternative energy sources claimed capable of replacing fossil fuels. Solar technology, hydrogen technology, and fusion don't yield as much emergy as they use. Other sources are too limited in quantity. The economy will use solar energy in agrarian production to its fullest, but this will only support about 10 to 20 percent of developed countries' present emergy use. Barring something not now known, economic contraction and descent are certain. The chief obstacle to a smooth transition to a lower-energy economy is the belief—valid for centuries, but no more—that there will always be a new source of energy to replace dwindling current sources. Priority must go to electric power as the basis of information networks and medical care. Beyond the time of cheap oil and natural gas, the essentials of civilization will depend on renewable hydroelectric power and the remaining coal and nuclear fission resources.

To sustain any nation during transition, priority is needed to keep a primary fuel source with a high net emergy available to its own economy. Foreign exchange should be prioritized for fuel import because of its high net emergy contribution. Policies should arrange to use the real wealth of commodities and resources instead of selling them for the smaller emergy in the money. National control is needed for financial exchanges so as to limit interest lost to outside borrowing and ensure the retention and growth of capital at home. Policies should limit the emergy losses from financial exchange with more urban countries with lower emergy-money ratios. Laws should limit unnecessary cars and horsepower so that the fuels go into useful production.

The following emergy-based policies can maintain the quality of life for people during transition and downsizing. The maximum money motivation should be replaced with the *maximum empower ethic*. An upper limit should be placed on income for individuals so that profits will go into reinvestment in production instead of affluence and waste. A universal policy should be developed that downsizes by cutting salaries rather than by firing people. More people should be kept at work with age-based, lower wage scales for teenagers and the elderly. Use of a permanent, minimum-wage public works program would assure full employment and help with necessary changes in infrastructure. A *universal baseline health care system* to include preventative medicine, pharmaceuticals, pregnancy care, and emergency treatments should be developed, but experimental medicine, luxury treatments, and long hospitalizations should be left to private insurance.

Polices for Descent

As soon as it is clear that the stage of descent has begun, urgent efforts will be needed to reduce population at the same rate as the annual emergy becomes less available. Money supplies should be adjusted to emergy use to prevent inflation or deflation. The ratio of purchased emergy to local emergy should be used to estimate the appropriate intensity of activities and their resource uses for the stages ahead. For a time of negative growth, we must expect new ways to finance reorganization with less borrowing and less use of stocks and bonds. Whereas government costs have to decrease, government roles in stabilizing descent may have to increase.

With less fuel use and fewer automobiles, cities will decentralize and more people will disperse to towns and farms. Providing population is reduced, the transition offers an opportunity to solve many critical urban problems, replace infrastructures, reintroduce green areas, cluster living closer to jobs, reduce commuting, dismantle strip developments and excess advertising, gradually develop more permanent housing structures, replace many roads with bikeways, and let information functions and their workers redevelop the central city.

To maximize the contributions of the hydrologic cycle, allowing nature's processes to reorganize patterns of streams, estuaries, and beaches will maximize the work of water. Without the cheap fuels to justify continued water diversions, floodplain levees, locks, reservoirs, nonrenewable uses of groundwater, and coastal groundwater use should be eliminated. Dams should be limited except where electric power for information has priority. All dilute wastewater should be routed through wetlands rather than into streams and estuaries. Overfishing of public waters should be limited and low-intensity aquaculture used as a substitute.

For the new conditions, productivity of lands should be maximized by developing a mosaic of land uses, each managed with appropriate cycles. The mosaic should include areas of lower-intensity agriculture, strips of complex ecosystems capable of making agricultural land rotation efficient while conserving high-diversity gene pools, areas for forestry plantations, areas for food-producing orchards and household gardens, and wetland areas for water management. Aid should be provided for the gradual substitution of fuel-using machinery, pesticide, and fertilizer by human labor and organic practices in the areas producing food, fiber, and wood.

Whereas not all the information and knowledge from the zenith of civilization can be maintained in an active state when resources are scarce, universal sharing of essential information is the great hope for a less-intense, smaller-but-prosperous economy. Knowledge should be divided into two categories: (1) essential knowledge for operating during descent, and (2) inactive knowledge to be preserved by information-conservation institutions. Universal understanding is needed of the changes and new ideals appropriate for descent. In spite of decentralization, a global network of common ideals and respect needs to be retained as part of basic education. As a society's institution for new ideas and plans, leading universities need to make the descent mission—including the consolidation of knowledge and changes in education—a primary purpose. Providing population is reduced, society can afford to educate all preschool- and school-age children with the money saved from operating fewer prisons.

The measures on several scales will be successful if they provide each person a job, baseline health care, an efficient lifestyle, a place in some kind of extended family, a cooperative relationship with diverse peoples, and a belief in the importance of his or her own opportunity to contribute to the economy and environment.

Contraction rather than expansion needs different measures, especially the belief that a decentralized civilization can be a better place to live. We end this book asserting a long-range ideal: *All members of society must preserve knowledge, sustain progress, and serve the Earth in ways appropriate to the stage of its cycle.*

Epilogue, the Distant Future

The general ecological model for self-organizing systems helps us think about the distant future, the fourth stage in the universal pulsing cycle (Figure 5.2). After the period of coming down, a time of Earth restoration may follow. Before another cycle of growth, consumption, and highly developed civilization can start, there has to be a restoration of the slowly renewable resources: the forests, the soils, the fisheries, and some of the wealth from geological processes. During this era the human economy has to operate on renewable resources: the sun, wind, rain, waves, tides, and geological processes, but in a restrained enough way for the biosphere to recover its natural capital and productivity. By this time population should be small enough to maintain high standards of living, as evaluated in measures of real wealth.

Information of the Roman, Greek, and Eastern civilizations was kept alive during later quiescent periods by low-energy institutions specializing in knowledge. Information stored from past glory became sacred, and human culture was programmed to become keepers of past knowledge. Perhaps, this time, keeping of past knowledge will be with a longer and larger perspective, understanding that more growth periods are to follow.

Stewardship for the Period of Restoration

If the Earth is to maximize its performance through the full cycle, a human culture will be needed that helps environmental restoration. We may expect ethics to develop in which people collectively aid net production processes that restore the resource reserves, the soils, the forests, the peat deposits, and the mineral deposits moving slowly up from below ground in the Earth cycles. As sometimes attributed to past cultures, people may find glory in being an agent of the Earth. It remains to be seen whether the social mechanisms will be conscious, logical, emotional, ritualistic, regimented, or by some means that we can't imagine.

NOTES

Chapter 1: Introduction to the Way Down

1. The concept of *emergy* was introduced in many publications in scientific literature starting in 1983. A thing's *emergy* refers to the amount of one kind of energy previously used to make it, whereas its energy is still contained within that thing. Definitions and explanations are given in Chapter 4. *Environmental Accounting, Emergy and Decision Making* (H. T. Odum, 1996) reviews the published literature. It contains many of the quantitative evaluations of emergy discussed in this book.
2. H. K. Okruszko, 1985.
3. To get systems diagrams like Figures 4.2 and 5.5 in mind, examine the outside sources from which energy comes as forcing functions. Trace flows along pathways. Scan the storages (*tank symbols*) and consider the effects of accumulations there. Visualize forces along the pathways that come from storages and outside sources. Trace pathways that go to *intersection symbols*, which are the places of production. Note the pathways of used energy dispersing downward and out of the system through the *heat sink symbol*. Because the diagrams are drawn with items from left to right according to the concept of energy hierarchy (see Chapter 4), think about the left-right series of energy transformations in the web of symbols. Look for the feedback pathways that go from right to left and cause some reinforcement of process there. Look for a pathway of recycle from right to left. Think about the effects of adding a surge of something.
4. An example of events galvanizing public change is the retreat from intensive beef agriculture caused by the spread of the "mad cow disease" in Europe in 2001. Events that change society toward what is sustainable in the next era are likely to become permanent. In this case, over-energized agriculture has to make a better fit with humanity and environment.

Chapter 2: The Present Condition

1. A. B. Robinson, S. L. Baliunas, W. Soon, and Z. W. Robinson 1998; J. Oerlemans 1994.
2. Graph given by L. R. Brown and C. Flavin 1999.
3. Data reported by L. R. Brown et al. 1999; L. R. Brown, H. Kane, and D. M. Roodman 1994; W. Lutz 1996.
4. See Figure 2.3.
5. See Figure 11.1.
6. Data from U.S. Statistical Abstracts.
7. Modified from C. Flavin and S. Dunn 1999.
8. Trend graph showing declining ratio of tons carbon to tons of oil energy by N. Nakićenović, et al. 1996.
9. B. Fleay 1995; H. Daly 1996; W. Youngquist 1997; Brandenburg and Paxson 1999; M. J. Economides 2000; status of global energy uses and efficiencies is summarized by V. Smil 1999; W. Fey and A. Lam 2000.
10. J. E. Young 1992; J. E. Young and A. Sachs 1994.
11. Cropland loss: G. Gardner 1996; irrigated land damaged: S. Postel 1989.
12. From R. H. Waring and S. W. Running 1998: Forest: 52.6 million square kilometers = 13 billion acres. World wood consumption from A. B. Durning 1993.
13. Coral reefs: A. P. McGinn 1999; environmental treaties: H. F. French 1995.
14. L. R. Brown, H. Kane, and D. Roodman 1994; L. R. Brown, C. Flavin, and H. French 2000.
15. Wind increases: L. R. Brown, H. Kane, and D. Roodman 1994.
16. U.S. Statistical Abstract for 1998.
17. H. Kane 1995.
18. H. T. Odum 1999.
19. Vital signs: L. R. Brown, H. Kane, and D. Roodman 1994.
20. Military expenditures: M. Renner 1989; arms sales: H. F. French 1993.
21. Increased paper use: L. R. Brown, H. Kane, and D. Roodman 1994; information technology growth: Figure 2.12; nature of computer growth: K. Kelley 1998.
22. Electric power: C. Flavin and N. Lenssen 1994.
23. Estimate of radioactive wastes from N. Lenssen 1991.
24. Alan Greenspan, chairman of the U.S. Federal Reserve Board, made two public presentations in 1999 describing stocks as overpriced.
25. Increasing trade: H. F. French 1993; exports: A. B. Durning 1989.
26. Debt: L. R. Brown, H. Kane, and D. Roodman 1994.

Chapter 4: The Ways of Energy and Materials in All Systems

1. H. T. Odum 1996.
2. H. T. Odum and R. C. Pinkerton 1955.

Chapter 5: Pulsing and the Growth Cycle

1. After earlier studies of oscillating population models, C. S. Holling (1986) used a now well-known "figure 8" diagram to generalize about "nature's cycles." See review in Chapter 3 of this book. H. T. Odum (1999) explains the relation of Holling's "figure 8" model to the concept of energy hierarchy. Previous statements of the pulsing paradigm were given in the last chapter in H. T. Odum 1983 and W. P. Odum, E. P. Odum, and H. T. Odum, 1995.

2. The model in Figure 5.3 generates the oscillations in Figure 5.2 (J. F. Alexander 1978). It has the following equations:

   $R = J/(1 + k_0{}^*M{}^*Q)$, the remaining unused energy in the source inflow.

   $M = M_t - 0.1{}^*Q$, the quantity of dispersed materials available.

   $dQ/dt = k_1{}^*R{}^*M{}^*Q - k_2{}^*Q - k_3{}^*Q{}^*A{}^*A$, the rate of change of resource reserve.

   $dA/dt = + k_4{}^*Q{}^*A{}^*A + k_5{}^*Q - k_6{}^*A$, the rate of change of consumer assets, where J is the energy source inflow, M the storage of unused materials, Q the resource reserve, M_t the total materials in the whole system, and A the consumer assets.

   The following terms of the equations go with the symbols in Figure 5.3. For the pointed block symbol on the left the net output is: $k_1{}^*R{}^*M{}^*Q$. For the second pointed block symbol to the right the net output is: $k_4{}^*Q{}^*A{}^*A$. For the pathway with the small block the output is: $k_5{}^*Q$.

3. The recent Ph.D. dissertation by D. Kang (1998) has extensive simulations of pulsing models and their energetics alone and in various combinations.

4. M. Harris 1977.

Chapter 6: Real Wealth and the Economy

1. The simulation model *Miniwrld* in Figure 6.8 is the model in Figure 5.5a and Chapter 5 Note 2 with a constant rate of circulating money and prices added. The following are the lines in the simulation program in BASIC.

```
1 REM PC
3 REM MINIWRLD
5 CLS
8 SCREEN 1,0
12 COLOR 0,1
17 LINE (0,0)-(320,300),3,B
20 LINE (0,150)-(320,150),3
25 REM Starting Values
27 F = 150
```

30 Q = 31.5
35 REM Scaling factors
40 Q0 = 1.3
45 DT = .1
50 T0 = .18
55 P0 = 2
60 F0 = 1.3
65 pr0 = .3
67 REM Renewable source
75 J = 3
80 REM coefficients
85 K1 = .002
90 K2 = .1
95 K3 = .001
100 REM FUEL RESERVES
105 PSET (T/T0, 150 - F/F0),3
110 REM WORLD ASSETS
120 PSET (T/T0,150-Q/Q0)
125 REM Prices
130 PSET (T/T0, (300-pr/pr0)),3
135 P = K1*F*Q*X +J
140 pr = 100/P+.03
145 DF = - K3 * F * Q*X
150 DQ = K1 * F * Q*X + J - K2 * Q
155 F = F + DT* DF
160 Q = Q + DT * DQ
165 T = T + DT
175 IF T/T0 >50 THEN X=1
180 IF T/T0 < 320 GOTO 100

Chapter 7: Spatial Organization

1. G. Stanhill (1977) quantitatively evaluated the system of highly productive agriculture supporting Paris, France, in the last century, with horses, transportation, and the recycling of horse manure to the land.
2. The measure of concentration of emergy use is "areal empower density," which can be represented in units of solar emcalories per acre per year.
3. Maps of empower density and transformity were developed by D. F. Whitfield (1994) for Jacksonville and S. L. Huang (1998) for Taipei, Taiwan.
4. W. Christaller (1933) recognized spatial hierarchy with hexagonal polygons of support and influence around villages and towns in agrarian landscapes of Europe before the industrial revolution. We observed them from an airplane near New Delhi, India, in 1970.
5. U. Sundberg et al. (1994) made an emergy evaluation of the Swedish empire of the seventeenth century with its economic and military power based on the emergy of its forest and mineral resources.

Chapter 8: Population and Wealth

1. The equation for the rate of change of the population in Figure 8.1 is:

$$dN/dt = L_1*N*(A/N) - k_7*N*(1-k_9*A) - k_8*N*N*(1 - k_9*A)$$

    where N is the population number, A is economic assets, and the ks are coefficients. Birth rate is $L_1*N*(A/N)$. Ordinary death rate is $k_7*N*(1 - k_9*A)$, and death rate from epidemics is $k_8*N*N*(1 - k_9*A)$ where $(1 - k_9*A)$ is the reduction of mortality due to public health and medicine.

2. Equations for the effect of population on the production process in Figure 8.2a follow from the symbols where J is the inflow of emergy from sources, R is the unused flow of source emergy, N is the population number, and P is empower production.

    $$R = J - k_0*R*N \text{ and therefore } R = J/(1 + k_0*N)$$

    $P = k_1 *R*N$, which is the equation for the graph in Figure 8.2b

3. R. J. Beyers and H. T. Odum (1993) reviewed the scientific studies on supporting humans in closed systems. A special issue of *Ecological Engineering* carried the research papers on Biosphere 2 (B. Marino and H. T. Odum 1999).

4. The population model in Figure 8.4 is a modification on one by H. T. Odum and Graeme Scott that has two production functions, one based only on renewable resource inflow and one based on the product interaction of nonrenewable resources available, the unused renewables, and the population available to work. The population reproduction (population production) is a product of economic assets per person and the population number (in which the population number cancels out). Two mortality flows are included, one the regular linear one in proportion to population, and an epidemic flow that is in proportion to the square of the population (representing contagious infection). Both are diminished in proportion to the economic assets available for public health prevention and treatment. The equations are:

    $$dF/dt = - k_0*R*F*A*N$$

    $$R = J/(1 + k_1*F*N*A + k_2*A)$$

    $$dA/dt = k_3*R*F*N*A + k_4*R*A - k_5*A - L_0*N*N*(1 - k_9*A)$$
    $$- L_0*N*(1 - k_9*A) - k_6 *A*N/N$$

    $$dN/dt = L_1 A - k_7*N*(1 - k_9*A) - k_8*N*N*(1 - k_9*A)$$

    where F is the nonrenewable reserve, J the inflow of renewable resources, R the unused flow of renewable resources, A economic assets, and N the population number. The program PEOPLE in BASIC was listed elsewhere (H. T. Odum and E. C. Odum 2000).

Chapter 9: The Global Network

1. H. T. Odum 1991; H. T. Odum and E. C. Odum 2000.

2. C.A.S. Hall 1999.
3. Emdollars were calculated as the emergy divided by the U.S. emergy-money ratio. For example 7.6 x 10^{11} solar emcalories of Japanese real wealth divided by 1.7 x 10^{11} solar emcalories per international dollar is 4.5 emdollars.
4. Emergy of payback was calculated by multiplying $1.10 by 2.0 x 10^{12} solar emcalories per dollar from Table 6.1.
5. M. T. Brown et al. 1992.
6. P. Dalton 1987; See H. T. Odum 1996.
7. U. Sundberg et al. 1994.

Chapter 10: Energy Sources
1. At full power, the efficiency of converting heat into the mechanical energy of machinery is given by the following:

   Efficiency = (1/2) (DT/T) where T is the Kelvin temperature (Celsius temperature + 273) and DT is the temperature difference between the intake and environment of the heat engine.

2. C.A.S. Hall, C. J. Cleveland, and R. Kaufmann 1986.
3. C. Lapp 1991.
4. S. J. Doherty 1995.
5. Ibid.
6. Ibid.
7. Net emergy calculations were provided from H. T. Odum 1996.
8. Emergy evaluation of two solar-voltaic systems are given by H. T. Odum 1996. One of these is the analysis of the power grid at Austin, Texas, by Robert King.
9. Living photoelectric cells made by putting blue-green algal mats in tubes (N. E. Armstrong and H. T. Odum 1963).
10. It is a property of energy transformation that the more you load a device to convert energy the lower the efficiency. You can convert a small amount of energy at high efficiency or a lot of energy at low efficiency. For example, solar-voltaic cells (green plants and human-made devices) convert 50 percent or more at low light intensity, but convert only 1–5 percent of sunlight at high intensity. This inverse relation between efficiency and amount of energy is due to the basic nature of energy thermodynamics. Humans can do nothing to change this relationship. Many misleading projections have been offered for solar energy by people extrapolating low light efficiencies to estimate power yields of daylight.

Chapter 11: Sustaining a Nation
1. Details on the structure and organization of ecosystems are given in Chapters 21–22 in H. T. Odum 1983, 1994.
2. M. J. Canoy 1972.

Chapter 12: Sustaining People

1. G. Hardin (1993) relates the history of alternative, drastic means of population regulation. M. Harris (1977) described the Aztec system of sacrificing 100,000 prisoners in a ceremony as a form of population regulation.

Chapter 13: Starting Down

1. H. T. Odum ended his 1971 book *Environment, Power, and Society* with the question: "Prophet where art thou?" *Prophet* was used in the sense of a person who had gained the faith and attention of all the people. Although the emergy principles gave intellectuals some vision of the future, even in 1971 it was clear that except among some of their students, the universities had lost the center stage of ideas to the media, who were mostly concerned with growth ethics and the short range.
2. A suggestion by Robert King, Good Company, Austin, Texas.

Chapter 14: Reorganizing Cities

1. Hierarchical organization of human movement in cities was recognized by C. A. Doxiadis (1968) as a basis for planning transportation. He found many small-scale movements in local areas and fewer large-scale movements over larger areas to and from the centers of cities.
2. Study by B. Van Der Loop given in H. T. Odum, E. C. Odum, and M. Blissett 1987.
3. Lowe, et al. 1997.
4. E. Jokela and W. H. Smith 1990.
5. The use of wetlands to recondition wastewaters that started with a project of the Rockefeller Foundation and National Science Foundation at the University of Florida in Gainesville, Florida, in 1973 (H. T. Odum et al. 1977; K. C. Ewel and H. T. Odum 1978, 1985; R. Kadlec and R. L. Knight 1998) has now become a general procedure worldwide. Also see the results of studies of wetlands receiving heavy metals (H. T. Odum et al. 2000).

Chapter 15: Restoring Waters

1. Based on an overview simulation model relating global carbon dioxide, fossil fuel use, deforestation, and oceanic carbonates (H. T. Odum 1995; H. T. Odum and E. C. Odum 2000).
2. J. N. Kremer and J. S. Lansing 1995.
3. Wetland research, note 4 for Chapter 14, above.
4. C. Diamond 1984; H. T. Odum, C. Diamond, and M. T. Brown 1987.
5. See Endnote 3.
6. K. C. Ewel and H. T. Odum 1985.
7. Ibid.
8. H. T. Odum and J. E. Arding 1991.
9. H. T. Odum 1996.

Chapter 16: Refreshing the Landscape

1. H. T. Odum 1996, p. 186.
2. Late summary of tropical forestry (F. H. Wadsworth 1998).
3. C.A.S. Hall, C. J. Cleveland, and R. Kaufmann 1986; H. T. Odum et al. 1976.
4. See summary of concepts and examples in W. J. Mitsch and S. E. Jorgensen 1989.

Chapter 17: Transmitting Knowledge

1. The authors have written a text for this course. See H. T. Odum, E. C. Odum, and M. T. Brown 1997.
2. H. R. Clinton 1996.

Chapter 18: Preparing People

1. W. C. Allee 1938, 1951.

REFERENCES CITED

Adams, B. 1986. *The Law of Civilization and Decay: An Essay on History.* MacMillan, New York, 393 pp.

Adams, R. N. 1988. *The Eighth Day: Social Evolution as the Self Organization of Energy.* Univ. of Texas Press, Austin, 282 pp.

Alexander, J. F., Jr. 1978. "Energy Basis of Disasters and Cycles of Order and Disorder." Ph.D. Dissertation, Enviromental Engineering Sciences, Univ. of Florida, Gainesville, 232 pp.

Allee, W. C. 1938, 1951. *Cooperation Among Animals.* Henry Schuman, New York, 233 pp.

Armstrong, N. E., and H. T. Odum. 1963. "Photoelectric Ecosystem." *Science* 143: 256–258.

Barber, B. 1996. *Jihad vs. McWorld: How Globalization and Tribalism Are Reshaping the World.* Random House, New York, 389 pp.

Barnett, R. J. 1980. *The Lean Years: Politics in the Age of Scarcity.* Simon and Schuster, New York, 347 pp.

Barney, G. O. 1980. *Global 2000 Report to the President,* V. 3. U.S. Government Printing Office, Washington, D.C., 401 pp.

Beckman, R. 1983. *The Downwave: Surviving the Second Great Depression.* Milestone Publications, Portsmouth, Great Britain, 363 pp.

Bell, A. 1997. *The Quickening: Today's Trends, Tomorrow's World.* Paper Chase Press, New Orleans, 333 pp.

Bell, D. 1973. *The Coming of Post-Industrial Society.* Basic Books, New York, 507 pp.

Bell, D. 1976, 1998. *The Cultural Contradictions of Capitalism.* Basic Books, New York, 363 pp.

Bell, D. 1995. "Introduction: Reflections on the End of an Age." In *Encylopedia of the Future,* ed. G. T. Kurian and G.T.T. Molitor. MacMillan, New York.

Bennett, D. J. 1998. *Randomness.* Harvard Univ. Press, Cambridge, MA, 238 pp.

Beyers, R. J., and H. T. Odum. 1993. *Ecological Microcosms.* Springer-Verlag, New York, 555 pp.

Brandenburg, Dr. J. E., M. R. Paxson, and J. E. Brandenburg. 1999. *Dead Mars, Dying Earth*. Crossing Press, Freedom, CA, 376 pp.

Brown, L. R., ed. 2000. *State of the World*. Worldwatch Institute, Washington, D.C.

Brown, L. R., C. Flavin, and H. French. 1999. *State of the World 1999*. W. W. Norton, New York, 259 pp.

Brown, L. R., C. Flavin, J. N. Abramovitz, S. Dunn, G. Gardner, A. R. Mattoon, A. P. McGinn, M. O'Meara, M. Renner, D. Roodman, P. Sampat, L. Starke, and J. Tuxill. 1999. *State of the World 1999, Millennial Edition*. W. W. Norton Co. and Worldwatch Institute, Washington, D.C.

Brown, L. R., and C. Flavin. 1999. Chapter 1, "A New Economy for a New Century," in Brown et al. 1999.

Brown, L. R., C. Flavin, and H. French. 2000. *State of the World 2000*. Worldwatch Institute, W. W. Norton, New York, 276 pp.

Brown, L. R., H. Kane, and D. M. Roodman. 1994. *Vital Signs 1994*. Worldwatch Institute, W. W. Norton, New York, 158 pp.

Brown, L. R., M. Renner, and B. Halweil. 1999. *Vital Signs 1999*. Worldwatch Institute, W. W. Norton, New York, 198 pp.

Brown, M. T., T. P. Green, A. Gonzalez, and J. Venegas. 1992. *Emergy Analysis Perspectives, Public Policy Options and Development Guidelines for the Coastal Zone of Nayarit, Mexico*. Center for Wetlands, Univ. of Florida, Gainesville, Vol. 1, 215 pp.; Vol. 2, 145 pp. and 31 map inserts.

Brown, M. T., and S. Ulgiati. 1999. "Emergy Evaluation of the Biosphere and Natural Capital." *Ambio* (in press).

Campbell, C. J. 1997. "Depletion Patterns Show Change Due for Production of Conventional Oil." *Oil and Gas J.* (Dec.): 33–39.

Campbell, C. J. 1997. *The Coming Oil Crisis*. Multi-Science Publishing Company, Brentwood, Essex, England, 210 pp.

Canoy, M. J. 1972. "Deoxyribonucleic Acid in Ecosystems." Ph.D. Dissertation, Univ. of North Carolina, Chapel Hill, 298 pp.

Capra, F. 1982. *The Turning Point: Science, Society, and the Rising Culture*. Simon and Shuster, New York, 464 pp.

Catton, W. R. 1982. *Overshoot*. Univ. of Illinois Press, Champaign, 298 pp.

Celente, G. 1997. *Trends: How to Prepare for and Profit from Changes of the 21st Century*. Warner Books, New York, 337 pp.

Cetron, M., and T. O'Toole. 1982. *Encounters with the Future: A Forecast of Life into the 21st Century*. McGraw Hill, New York, 308 pp.

Christaller, W. 1933, 1966. *Central Places in South Germany*, trans. C. W. Baskin. Prentice Hall, Englewood Cliffs, NJ, 230 pp.

Clinton, H. R. 1996. *It Takes a Village: And Other Lessons Children Teach Us*. Simon and Schuster, New York, 336 pp.

Coates, J. F., and J. Jarret, eds. 1996. The *Future: Trends into the Twenty-first Century*. Sage, Newbury Park, CA, 198 pp.

Collinvaux, P. 1980. *The Fates of Nations: A Biological Theory of History*. Simon and Schuster, New York, 383 pp.

Commoner, B. 1979. *The Politics of Energy*. A. E. Knopf, New York, 101 pp.

Cook, E. F. 1976a. "National Energy Future, Problems and Policy Issues." In *Middle and Long-term Energy Policies and Alternatives*, Pt. 1, Subcommittee on Energy and Power, Committee on Interstate and Foreign Commerce, House of Representatives, U.S. 94th Congress. Serial No. 94-63, U.S. Government Printing Office, Washington, D.C., 613 pp.

Cook, E. F. 1976b. *Man, Energy, and Society.* W. H. Freeman and Co., San Francisco, 491 pp.

Corning, P. 1983. *The Synergism Hypothesis: A Theory of Progressive Evolution.* McGraw Hill, New York, 491 pp.

Costanza, R., ed. 1991. *Ecological Economics: The Science and Management of Sustainability.* Columbia Univ. Press, New York, 525 pp.

Cottrell, W. F. 1955. *Energy and Society.* McGraw Hill, New York, 330 pp.

Cox, W. M., and R. Alm. 1999. *Myths of Rich and Poor: Why We're Better Off Than We Think.* Basic Books, New York, 256 pp.

Dalton, P. 1987. "Defense Analysis." Student policy research project report, Lyndon B. Johnson School of Public Affairs, Univ. of Texas, Austin, 14 pp.

Daly, H. 1996. *Beyond Growth.* Beacon Press, Boston, 253 pp.

Davis, B., and D. Wessel. 1998. *Prosperity: The Coming Twenty-Year Boom and What It Means to You.* Times Business, New York, 324 pp.

Davis, W. J. 1979. *The Seventh Year: Industrial Civilization in Transition.* W. W. Norton, New York, 296 pp.

Dent, H. S. 1998. *The Roaring 2000s: Building the Wealth and Lifestyle in the Greatest Boom in History.* Simon and Schuster, New York, 319 pp.

Dent, H. S. 1994. *The Great Boom Ahead.* Hyperion Press, Westport, CT, 272 pp.

Diamond, C. 1984. "Energy Basis for the Regional Organization of the Mississippi River Basin." M.S. Thesis, Environmental Engineeering Sciences, Univ. of Florida, Gainesville, 136 pp.

Diamond, J. 1998. *Guns, Germs, and Steel: The Fates of Human Societies.* W. W. Norton, New York, 480 pp.

Dobkowski, M. N. and I. Wallimann, eds. 1998. *The Coming Age of Scarcity: Preventing Mass Death and Genocide in the Twenty-first Century.* Syracuse Univ. Press, Syracuse, New York, 352 pp.

Doherty, S. J. 1995. "Emergy Evaluations of and Limits to Forest Production." Ph.D. Dissertation, Environmental Engineering Sciences, Univ. of Florida, Gainesville, 203 pp.

Doxiadis, C. A. 1968. "Man's Movement and His City." *Science* 162: 326–332.

Duncan, R. C. 1993. "The Life-Expectancy of Industrial Civilization, the Decline to Global Equilibrium." *Population and Environment,* 14: 325–357.

Durning, A. B. 1989. "Poverty and the Environment: Reversing the Downward Spiral." Worldwatch Institute Paper #92. Washington, D.C., 86 pp.

Durning, A. B. 1992. *How Much Is Enough?* W. W. Norton, New York, 200 pp.

Durning, A. B. 1993. "Saving the Forests: What Will It Take?" Worldwatch Institute Paper #117. Washington, D.C., 51 pp.

Economides, M. 2000. *The Color of Oil.* Round Oak Publ. Co., Katey, Texas, 220 pp.

Ehrlich, P. R. 1968. *The Population Bomb.* Ballentine, New York, 223 pp.

Ehrlich, P. R. 1974. *The End of Affluence.* Ballentine, New York, 304 pp.

Elgin, D. 1981. *Voluntary Simplicity.* William Morrow & Co., New York, 312 pp.

Ewel, K. C., and H. T. Odum. 1978. "Cypress Swamps for Nutrient Removal and Wastewater Recycling." In *Advances in Water and Wastewater Treatment: Biological Nutrient Removal,* ed. M. P. Wanelista and W. W. Edkenfelder, Jr. Ann Arbor Sci. Publ. Inc., Ann Arbor, MI.

Ewel, K. C., and H. T. Odum, eds. 1985. *Cypress Swamps.* Univ. of Florida Press, Gainesville, 472 pp.

Fernald, E. A., E. D. Purdum, J. R. Anderson, Jr., and P. A. Kroft. 1992. *Atlas of Florida.* Univ. Press of Florida, Gainesville, 280 pp.

Flavin, C., and S. Dunn. 1999. "Reinventing the Energy Systems." In Brown et al. 1999.

Flavin, C., and N. Lenssen. 1994. *Power Surge: Guide to the Coming Energy Revolution.* Worldwatch Environmental Alert Series, W. W. Norton, New York, 382 pp.

Flavin, C., and N. Lenssen. 1994. "Powering the Future: Blueprint for a Sustainable Electricity Industry." Worldwatch Institute Paper #119. Washington, D.C., 74 pp.

Flavin, C., and O. Tunali. 1996. "Climate of Hope: New Strategies for Stabilizing the World Atmosphere." Worldwatch Institute Paper #130. Washington, D.C., 84 pp.

Fleay, B. 1995. *The Decline of the Age of Oil, Petrol Politics: Australia's Road Ahead.* Pluto Press, London, U.K., 152 pp.

Forrester, J. W., and N. J. Mass. 1976. "U.S. Long-term Energy Policy in a Changing National Environment." In *Middle and Long-Term Energy Policies and Alternatives,* Pt. 1, Subcommittee on Energy and Power, Committee on Interstate and Foreign Commerce, House of Representatives, U.S. 94th Congress, Serial No. 94-63, U.S. Government Printing Office, Washington, D.C., 613 pp.

Frank, A. G. 1998. *Reorient: Global Economy in the Asian Age.* Univ. of California Press, Berkeley.

Frank, A. G., and B. K. Gills. 1996. *The World System: Five Hundred Years or Five Thousand.* Routledge, New York, 352 pp.

Freeman, D. 1974. *Energy: The New Era.* Vintage Books, Random House, New York, 386 pp.

French, H. F. 1993. "Costly Tradeoffs: Reconciling Trade and the Environment." Worldwatch Institute Paper #113. Washington D.C., 74 pp.

French, H. F. 1995. "Partnership for the Planet: An Environmenal Agenda for the United Nations." Worldwatch Institute Paper #126. Washington, D.C., 71 pp.

Fukuyama, F. 1992. *The End of History and the Last Man.* The Free Press, New York, 418 pp.

Fuller, R. B. 1968. "An Operating Manual for Space Ship Earth." In *Environment and Change: The Next Fifty Years,* ed. W. R. Ewald, Jr. Indiana Univ. Press, Bloomington.

Fuller, R. B. 1973. *Earth, Inc.* Anchor Books, Garden City, New York, 180 pp.

Gardner, G. 1996. "Shrinking Fields: Cropland Loss in a World of Eight Billion." Worldwatch Institute Paper #131. Washington, D.C., 55 pp.

Gates, J. 1998. *The Ownership Solution: Toward a Shared Capitalism for the 21st Century.* Addison-Wesley, Reading, MA, 388 pp.

Georgescu-Roegen, N. 1977. "The Steady State and Ecological Salvation: A Thermodynamic Analysis." *Bioscience* 27(4): 266–270.

Gever, J., R. Kaufmann, D. Skole, and C. Vorosmarty. 1991. *Beyond Oil: A Threat to Food and Fuel in the Coming Decades.* Ballinger Publ. Co. (Harper & Rowe), Cambridge, MA, 304 pp.

Gray, D. D., and W. F. Martin. 1975. *Growth and Its Implications for the Future.* Dinosaur Press (Readers Press), Branford, CT, 181 pp.

Greider, W. 1997. *One World Ready or Not: The Manic Logic of Global Capitalism.* Simon and Schuster, New York, 528 pp.

Hall, C.A.S., ed. 1995. *Maximum Power.* Univ. Press of Colorado, Niwot, 644 pp.

Hall, C.A.S. 2000. *Quantifying Sustainable Development: The Future of Tropical Economies.* Academic Press, San Diego, 761 pp.

Hall, C.A.S., C. J. Cleveland, and R. Kaufmann. 1986. *Energy and Resource Quality: The Ecology of the Economic Process.* John Wiley, New York, 577 pp.

Hardin, G. 1993. *Living within Limits: Ecology, Economics, and Population Taboos.* Oxford Univ. Press, New York, 339 pp.

Harris, M. 1977. *Cannibals and Kings.* Vintage, Random House, San Diego, 351 pp.

Harris, S. E. 1977. *The Death of Capital.* Pantheon Books, New York.

Hawken, P. 1994. *The Ecology of Commerce: A Declaration of Sustainability.* Harper Business, New York, 250 pp.

Heilbroner, R. 1978. *Beyond Boom and Crash.* Norton, New York, 111 pp.

Heilbroner, R. 1994. *21st Century Capitalism.* W. W. Norton, New York, 111 pp.

Henderson, H. 1981. *The Politics of the Solar Age: Alternatives to Economics.* Archer Press, Garden City, New York, 433 pp.

Henderson, H. 1991. *Paradigms in Progress: Life Beyond Economics.* Knowledge Systems, Indianapolis, IN, 293 pp.

Henderson, H. 1996. *Building a Win-Win World: Life Beyond Global Economic Warfare.* Berrett-Koehler, San Francisco, 398 pp.

Hine, T. 1991. *Facing Tomorrow: What the Future Has Been, What the Future Can Be.* A. E. Knopf, New York, 264 pp.

Holling, C. S. 1986. "The Resilience of Terrestrial Ecosystems: Local Surprise and Global Change." In *Sustainable Development of the Biosphere,* ed. W. C. Clark and R. E. Munn. Cambridge Univ. Press, New York.

Holling, C. S. 1995. "What Barriers, What Bridges." In *Barriers and Bridges to the Renewal of Ecosystems and Institutions,* ed. L. Gunderson, C. S. Holling, and S. Light. Columbia Univ. Press, New York.

Huang, S. L. 1998. "Spatial Hierarchy of Urban Energetic Systems." In *Advances in Energy Studies: Energy Flows in Ecology and Economy.* Proceedings of the International Workshop in Porto Venere, Italy, ed. S. Ulgiati. Musis, Museo della Scienza e dell'informazione Scientifica, Rome, Italy.

Hubbert, M. K. 1949. "Energy from Fossil Fuels." *Science* 109: 103–109.

Hubbert, M. K. 1969. "Energy Resources." In *Resources and Man.* National Academy of Sciences–National Research Council. W. H. Freeman, San Francisco.

Huntington, S. 1996. *The Clash of Civilizations and Remaking the World Order.* Simon & Schuster, New York.

Inhaber, H. 1982. *Energy Risk Assessment.* Gordon and Breach Science Publ., New York, 394 pp.

Jernigan, E. 1995. *America and the World 1995–2015.* Future Press, Ocala, FL, 102 pp.

Johnstone, I. M. 1978. "In Search of Steady State." Thesis for B.A. in Architecture, School of Architecture, Univ. of Auckland, New Zealand, 228 pp.

Jokela, E., and W. H. Smith. 1990. "Growth in Elemental Composition of Slash Pine after Sixteen Years after Treatment with Garbage Composted with Sewage Sludge." *J. Env. Quality* 19: 146–150.

Kadlec, R., and R. L. Knight. 1998. *Treatment Wetlands.* Lewis Publishers, Boca Raton, FL, 893 pp.

Kane, H. 1995. "The Hour of Departure: Forces that Create Refugees and Migrants." Worldwatch Institute Paper #125. Washington D.C., 56 pp.

Kang, D. 1998. "Pulsing and Self-Organization." Ph.D. Dissertation, Department of Environmental Engineering Sciences, Univ. of Florida, Gainesville, 283 pp.

Kaplan, R. 1994. "The Coming Anarchy." *Atlantic Monthly* (Feb.): 44–46.

Kaplan, R. 1996. *The Ends of the Earth.* Random House, New York, 476 pp.

Kelley, K. 1998. *New Rules for the New Economy: Ten Radical Strategies for a Connected World.* Viking, New York, 179 pp.

Kemp, T. 1990. *The Climax of Capitalism: The U.S. Economy in the 21st Century.* Longmans, New York, 251 pp.

Kennedy, P. 1993. *Preparing for the Twenty-first Century.* Random House, New York, 428 pp.

Kiplinger, K. 1998. *World Boom Ahead: Why Business and Consumers Will Prosper.* Kiplinger Books, Washington, D.C., 404 pp.

Knoke, W. 1997. *Bold New World: The Essential Road Map to the Twenty-first Century.* Kodansha International, New York, 368 pp.

Kremer, J. N., and J. S. Lansing. 1995. "Modeling Water Temples and Rice Irrigation in Bali: A Lesson in Socio-ecologic Communication." In *Maximum Power,* ed. C.A.S. Hall, Univ. Press of Colorado, Niwot.

Krishnan, R., J. M. Harris, and N. R. Goodwin. 1995. *A Survey of Ecological Economics.* Island Press, Washington, D.C., 384 pp.

Kuznets, S. 1930. *Secular Movements in Production and Prices.* Houghton & Mifflin, New York, 536 pp.

Lapp, C. 1991. "Emergy Analysis of the Nuclear Power System in the United States." Research paper for the M.E. degree. Environmental Engineering Sciences, Univ. of Florida, Gainesville, 64 pp.

Laszlo, E. 1974. *A Strategy for the Future.* George Brazilier, New York, 328 pp.

Laszlo, E. 1994. *Vision 2020: Reordering Chaos for Global Survival.* Gordon and Breach, Langhorne, PA, 133 pp.

Lenssen, N. 1991. "Nuclear Waste: The Problem that Won't Go Away." Worldwatch Institute Paper #106. Washington, D.C., 62 pp.

Lotka, A. J. 1922. "Contributions to the Energetics of Evolution: Natural Selection as a Physical Principle." *Proc. Natl. Acad. Sci.* 8: 147–154.

Lovins, A. B. 1977. *Soft Energy Pathways.* Ballinger, Cambridge, MA, 231 pp.

Lowe, E. A., J. L. Waren, and S. R. Moran. 1997. *Discovering Industrial Ecology, An Executive Briefing and Sourcebook.* Battelle Press, Columbus, Ohio, 191 pp.

Lutz, W. 1996. "The Human Race Slows to a Crawl." In *Options* (summer issue), International Institute of Applied Systems Analysis, Laxenburg, Austria.

Lyons, S. 1978. *Sun: A Handbook for the Solar Decade.* Friends of the Earth, San Francisco, 364 pp.

Madrick, J. 1995. *The End of Affluence.* Random House, New York, 223 pp.

Marino, B., and H. T. Odum, eds. 1999. *Biosphere 2: Research Past and Present.* Elsevier, New York, 259 pp.

McGinn, A. P. 1999. "Charting a New Course for Oceans." In Brown et al. 1999.

Meadows, D. H., D. L. Meadows, J. Randers, and W. W. Behrens III. 1972. *The Limits to Growth.* Potomac Associates Book, Washington, D.C., 205 pp.

Meadows, D. H., D. J. Richardson, and G. Bruckman. 1982. *Groping in the Dark: The First Decade of Global Modelling.* John Wiley, New York, 311 pp.

Meadows, D. H., D. L. Meadows, and J. Randers. 1992. *Beyond the Limits.* Chelsea Garden Publ., Post Mills, VT, 300 pp.

Milbrath, L. 1989. *Envisioning a Sustainable Society: Learning Our Way Out.* State Univ. Press of New York, Albany, 403 pp.

Miles, R. 1976. *Awakening from the American Dream: The Social and Political Limits to Growth.* Universe Books, New York, 246 pp.

Mills, S. 1997. *Turning Away from Technology.* Sierra Club, San Francisco, 256 pp.

Mitsch, W. J., and S. E. Jorgensen, eds. 1989. *Ecological Engineering: An Introduction to Ecotechnology.* John Wiley, New York, 472 pp.

Naisbitt, J. 1985. *Megatrends.* Warner Books, New York, 168 pp.

Naisbitt, J. 1996. *Global Paradox.* Avon Books, New York, 392 pp.

Nakićenović, N., P. V. Gilli, and R. Kurz. 1996. "Regional and Global Energy and Energy Efficiencies." *Energy* 21(3): 223–237.

Nussbaum, B. 1983. *The World after Oil: The Shifting Axiom of Power and Wealth.* Simon and Schuster, New York, 317 pp.

Odum, H. T. 1971. *Environment, Power and Society.* John Wiley, New York, 331 pp.

Odum, H. T. 1983. *Systems Ecology.* John Wiley, New York, 644 pp.

Odum, H. T. 1984. "Energy Analysis of the Environmental Role in Agriculture." In *Energy and Agriculture,* ed. G. Stanhill. Springer-Verlag, Berlin.

Odum, H. T. 1991. "Destruction and Power in General Systems." In *Proceedings of the 35th Annual Meeting of the International Society for the Systems Sciences,* Ostersund, Sweden, Vol. I, ed. S. C. Holmberg and K. Samuelson.

Odum, H. T. 1994. *Ecological and General Systems.* Univ. Press of Colorado, Niwot, 644 pp. (reprint of *Systems Ecology,* John Wiley, 1983)

Odum, H. T. 1995. "Tropical Forest Systems and the Human Economy." In *Tropical Forests: Management and Ecology,* Ecological Studies Vol. 112, ed. A. E. Lugo and C. Lowe. Springer-Verlag, New York.

Odum, H. T. 1996. *Environmental Accounting, Emergy and Decision Making.* John Wiley, New York, 370 pp.

Odum, H. T. 1999. "Limits of Information and Biodiversity." Pp. 230–269 in *Sozialpolitik und Okologieprobleme der Aukunft.* Austrian Academy of Sciences (Centennial), Vienna, Austria, 428 pp.

Odum, H. T., and J. E. Arding. 1991. "Emergy Analysis of Shrimp Maricul-
ture in Ecuador." Working paper prepared for the Coastal Resources
Center, Univ. of Rhode Island, Narragansett, 114 pp.

Odum, H. T., C. Diamond, and M. T. Brown. 1987. "Energy Systems Overview of the Mississippi River Basin." Report to The Cousteau Society.
Center for Wetlands, Univ. of Florida, Gainesville, 107 pp.

Odum, H. T., K. C. Ewel, W. J. Mitsch, and J. W. Ordway. 1977. "Recycling
Treated Sewage through Cypress Wetlands in Florida." In *Wastewater
Renovation and Reuse,* ed. F. D'Itri. Marcel Dekker, New York. (also printed
as Occasional Publ. #1, 1975, Center for Wetlands, Univ. of Florida)

Odum, H. T., C. Kylstra, J. Alexander, and N. Sipe. 1976. "Net Energy
Analysis of Alternatives for the United States." In Part 1. 94th Congress 2d Session Committee. Subcommittee on Energy and Power of
the Committee on Interstate and Foreign Commerce of the U.S. House
of Representatives, 94-63, U.S. Govt. Printing Office, Washington, D.C.,
613 pp.

Odum, H. T., and E. C. Odum. 1976, 1982. *Energy Basis for Man and Nature.*
McGraw Hill, New York, 337 pp.

Odum, H. T., and E. C. Odum. 2000. *Modeling for All Scales.* Academic Press,
San Diego, 458 pp.

Odum, H. T., E. C. Odum, and M. Blissett. 1987. "Ecology and Economy:
'Emergy' Analysis and Public Policy in Texas." Policy Research Project
Report #78. Lyndon B. Johnson School of Public Affairs, Univ. of Texas,
Austin, 178 pp.

Odum, H. T., E. C. Odum, and M. T. Brown. 1997. *Environment and Society in
Florida.* Lewis Publishers, Boca Raton, FL, 499 pp.

Odum, H. T., E. C. Odum, M. T. Brown, D. LaHart, C. Bersok, and J. Sendzimir.
1988. "Energy, Environment and Public Policy: A Guide to the Analysis of Systems." UNEP Regional Seas Reports and Studies No. 95. United
States Environment Programme, Nairobi, Kenya, 109 pp.

Odum, H. T., and R. C. Pinkerton. 1955. "Time's Speed Regulator: The
Optimum Efficiency for Maximum Power Output in Physical and Biological Systems." *Am. Scientist* 43(2): 331–343.

Odum, H. T., W. Wojcik, L. Pritchard, Jr., S. Ton, J. J. Delfino, M. Wojcik, S.
Leszczynski, J. D. Patel, S. J. Doherty, and J. Stasik. 2000. *Heavy Metals
in the Environment: Using Wetlands for Their Removal.* Lewis Publishers,
Boca Raton, FL, 325 pp.

Odum, W. P., E. P. Odum, and H. T. Odum. 1995. "Nature's Pulsing Paradigm." *Estuaries* 18(4): 547–555.

Oerlemans, J. 1994. "Quantifying Global Warming from the Retreat of
Glaciers." *Science* 264 (1994): 243–245. Graph quoted by Waring and
Running 1998.

Okruszko, H. K. 1985. "Decession in the Natural Evolution of Low
Peatlands." Pp. 89–94 in *Soil Ecology and Management,* ed. J. H. Cooley.
International Association for Ecology. Intecol Bulletin 1985: 12, Insti-
tute of Ecology, Athens, GA, 139 pp.

Ophuls, W. 1977. *Ecology and the Politics of Scarcity.* W. H. Freeman, San Franciso,
302 pp.

Ophuls, W., and A. S. Boyan, Jr. 1992. *Ecology and the Politics of Scarcity Revisited.* Freeman, New York, 379 pp.

Postel, S. 1989. "Water for Agriculture: Facing the Limits." Worldwatch Institute Paper #93. Washington, D.C., 54 pp.

Reich, R. B. 1983. *The Next American Frontier.* Penguin Books, New York, 324 pp.

Rees, W. E., and M. Wackernagel. 1994. "Ecological Footprints and Appropriated Carrying Capacity: Measuring the Natural Capital Requirements of the Human Economy." In *Investing in Natural Capital,* ed. A. M. Jansson, M. Hammer, C. Folke, and R. Costanza. Island Press, Washington, D.C.

Renner, M. 1989. "National Security: The Economic and Environmental Dimensions." Worldwatch Institute Paper #89. Washington, D.C., 78 pp.

Repetto, R. 1985. *The Global Possible: Resources, Development and the Next Century.* Yale Univ. Press, New Haven, CT, 538 pp.

Repetto, R. 1986. *World Enough and Time: Successful Strategies for Resource Management.* Yale Univ. Press, New Haven, CT, 147 pp.

Rifkin, J. 1995. *The End of Work: The Decline of the Global Labor Force and the Dawn of the Post-Market Era.* Tarcher/Putnam, New York.

Robertson, J. 1978. *The Sane Alternative: A Choice of Futures.* River Basin Publ. Co., St. Paul, 153 pp.

Robinson, A. B., S. L. Baliunas, W. Soon, and Z. W. Robinson. 1998. "Environmental Effects of Increased Atmospheric Carbon Dioxide." *Med. Sent.* 3(5): 171–178.

Rostow, W. W. 1998. *The Great Population Spike and After: Reflections on the 21st Century.* Oxford Univ. Press, New York, 256 pp.

Schumacher, E. F. 1973. *Small Is Beautiful: Economics as if People Mattered.* Harper and Row, New York, 324 pp.

Shumpeter, J. A. 1939. *Business Cycles.* McGraw Hill, New York, 461 pp.

Simon, J. L., and H. Kahn. 1984. *The Resourceful Earth: A Response to Global 2000.* Basic Blackwell, Oxford, England, 585 pp.

Simon, J. L. 1995. *The State of Humanity.* Blackwell Publ., Cambridge, MA, 694 pp.

Sinai, J. R. 1978. *The Decadence of the Modern World.* Schenkman Publ. Co., Cambridge, MA, 229 pp.

Smil, V. 1999. *Energies: An Illustrated Guide to the Biosphere and Civilization.* MIT Press, Cambridge, MA, 210 pp.

Soros, G. 1998. *The Crisis in Global Capitalism.* Perseus Books, Boulder, CO, 288 pp.

Stanhill, G. 1977. "An Urban Agroecosystem: The Example of 19th-Century Paris." *Agroecosystems* 3: 269–284.

Stobaugh, R., and D. Yergin, eds. 1979. *Energy Future: Report of the Energy Project of the Harvard Business School.* Ballantine Books, New York, 493 pp.

Strauss, W., and N. Howe. 1997. *The Fourth Turning: An American Prophecy.* Broadway Books, New York, 352 pp.

Sundberg, U., J. Lindegren, H. T. Odum, and S. J. Doherty. 1994. "Forest Emergy Basis for Swedish Power in the 17th Century." *Scandinavian Journal of Forest Research,* Supplement No. 1, 50 pp.

Tainter, J. 1988. *The Collapse of Complex Societies.* Cambridge Univ. Press, New York, 250 pp.

Tee, K. A. 1990. *Booms and Busts in Modern Societies.* Longman, Singapore, 260 pp.

Templeton, J. M. 1997. *Is Progress Speeding Up: Our Multiplying Multitudes of Blessing.* Templeton Foundation Press, Philadelphia, 291 pp.

Thompson, W. I. 1976. *Evil and World Order.* Harper and Row, New York, 116 pp.

Toffler, A. 1970. *Future Shock.* Random House, NY, 505 pp.

Toffler, A. 1980. *The Third Wave.* William Norcrop & Co., New York, 529 pp.

Toffler, A., and H. Toffler. 1994. *Creating a New Civilization.* Turner Publ., Atlanta, GA, 112 pp.

Toynbee, A. 1946. *A Study of History, Abridgement of Volumes I–VI,* by D.C. Somervell. Oxford Univ. Press, New York, 617 pp.

U.S. 94th Congress. 1976. Hearings of the Subcommittee on Energy and Power of the Committee on Interstate and Foreign Commerce, House of Representatives, Second Session on Energy Choices Facing the Nation and Their Long-range Implications, March 25–26, Serial No. 94-63, 607 pp.

U.S. Bureau of the Census. 1998. *Statistical Abstract of the United States 1998* (118th edition). National Technical Information Services, Washington D.C. 1019 pp.

Wadsworth, F. H. 1998. *Forest Production in Tropical America.* Handbook No. 710. U.S. Dept. of Agriculture, Forest Service, Washington, D.C., 563 pp.

Waring, R. H., and S. W. Running. 1998. *Forest Ecosystems Analysis at Multiple Scales,* 2d ed. Academic Press, San Diego, 370 pp.

Weber, P. 1994. "Net Loss: Fish, Jobs, and the Marine Environment." Worldwatch Institute Paper #120. Washington, D.C., 76 pp.

Whitfield, D. F. 1994. "Emergy Basis for Urban Land Use Patterns in Jacksonville, Florida." Master of Landscape Architecture Thesis, Univ. of Florida, Gainesville, 224 pp.

Wilhelm, R. 1968. *The I Ching.* Routledge & Kegan Paul, London, 806 pp.

Williams, P. 1973. *Das Energi.* Warner Books, New York, 149 pp.

Woithe, R. 1994. "Emergy Evaluation of the United States Civil War." Ph.D. Dissertation, Environmental Engineering Sciences, Univ. of Florida, Gainesville, 206 pp.

Wolman, W., and A. Colamosca. 1997. *The Judas Economy: The Triumph of Capital and the Betrayal of Work.* Addison-Wesley, Reading, PA, 240 pp.

Yergin, E., and J. Stanislaw. 1998. *The Commanding Heights: The Battle between Government and the Marketplace.* Simon and Schuster, New York, 464 pp.

Young, J. E. 1991. "Discarding the Throwaway Society." Worldwatch Institute Paper #101. Washington, D.C., 44 pp.

Young, J. E. 1992. "Mining the Earth." Worldwatch Institute Paper #109. Washington, D.C., 43 pp.

Young, J. E., and A. Sachs. 1994. "The Next Efficiency Revolution: Creating a Sustainable Materials Economy." Worldwatch Institute Paper #121. Washington, D.C., 58 pp.

Youngquist, W. 1997. *Geodestinies, the Inevitable Control of Earth Resources over Nations and Individuals.* National Book Co., Portland, OR, 500 pp.

INDEX